Fundamentals of Mass Transfer

Fundamentals of Mass Transfer

by

B. N. Nnolim, FNSChE, FNSE
Formerly of the
Department of Chemical Engineering
Nnamdi Azikiwe University, Awka

First Published September, 2008

Copyright © 2008 by Benedict Nnolim

ISBN 978-1-906914-01-1

All rights reserved. No part of this publication may be reproduced, stored in a retrieval system, or transmitted in any form or by any means, electronic, mechanical photocopying, recording or otherwise without the prior permission of the copyright owner.

Printed by Lulu
http://www.lulu.com

Ben Nnolim Books,
7 Sandway Path,
St. Mary Cray, Orpington,
Kent. BR5 3TS, UK.
benedictnnolim@aol.com

DEDICATION

This book is dedicated to the memory of Dr Henry Sawistowsky, Professor of Chemical Engineering, Imperial College of Science and Technology, London

Certainly the clearest, the best and the most liked Lecturer in our class of 1964-67.

PREFACE

Modern technology is going back to "nature" if the public concern, now in vogue, about the adverse effects on the environment by modern technology is anything to go by. It is, now, not only fashionable but also, could be, illegal if your commercial process was not "organic" or "environment friendly".

Indigenous technologies of the under developed societies, which are "organic" and "environment friendly" did not advance further after colonisation because they had very little of their philosophical background written down. They, also, had to compete with the apparently, more successful and conquering western technologies. Fortunately, they are dormant, not extinct.

Scientists, engineers and technologists from the developing countries do well when they study or work in the countries of the advanced world. They seem quite unable, however, when they work in their own countries, to solve the technical problems of their environment to any degree of satisfaction. There must be, obviously, some reasons for this state of affairs. My observation, however, seems to be that their technological culture is not detail oriented while the technological culture they must acquire to enter the mainstream of modern living is systematically detail oriented. It is almost the ancient argument, but with a twist, between Parmenides (Nothing changes), hence, according to the third world, no need to be concerned too much about things and Heraclitus (Everything changes), and, therefore, according to the developed world, must be understood and monitored.

There has been much progress since then. Parmenides and Heraclitus have been seen to be, in fact, on the right (steady state) and left (transient state) sides of the same curve. Not every scientist, engineer or technologist from the developing world will be trained or work in the developed world, however. They must acquire the systematic detail orientation in their own countries so that they can apply both the advanced western technologies as well as their indigenous technologies to the solution of their, and world, problems.

Basic chemical engineering subjects, such as mass transfer, heat transfer, momentum transfer, chemical reaction engineering and thermodynamics, no longer get the treatment in modern text books and journals that are useful or, indeed, helpful to new students in the subjects, especially, to students from developing countries where the technological culture is different. This is understandable since these books are directed, primarily, at the society of the advanced countries and have to deal with the level of technology and attitudes prevailing in those societies.

Although fundamentalism has, recently, been given a bad name in certain quarters, it is, still, essential to remind ourselves that the Pythagoras theorem, espoused thousands of years ago, is still valid today, that force is, still, equal to mass times acceleration and that since recorded history, we have been eating food through our mouths and will continue to do so for some time.

The late Professor Henry Sawistowsky, at Imperial College of Science and Technology London, gave such beautiful undergraduate and postgraduate lectures in Mass Transfer, often without reading from notes, of such clarity that it was a joy to attend his lectures. Several years later, I continue to be amazed at the clarity of his lectures. My amazement turned to action when I realised that students of, and practitioners in, mass transfer in developing countries do not have the opportunity to learn the basics. They have no choice but to skim and skip from one complex analysis to another without an appreciation of the fundamentals.

This book is, therefore, an opportunity for me to achieve two purposes. The first is to provide the opportunity for beginners in mass transfer, especially, those from developing countries to understand, fully, the basis of all the sophisticated analyses that is carried out in mass transfer. The second purpose is to honour, by so doing, the memory of Professor Sawistowsky, a brilliant teacher, by compiling and publishing what I heard from him. I have made only a few additions. Any shortcomings in this book should be blamed on me. I hope, however, that, inspite of my shortcomings, the brilliance and clarity of his lectures, given in the 1960s, will shine through.

I wish to acknowledge the support and encouragement given to me to carry out this project by his wife, Dr M. H. Sawistowka and the usefulness of the Wikipedia internet site for material in the appendices.

Benedict N. Nnolim
Professor of Chemical Engineering

TABLE OF CONTENTS

	DEDICATION	i
	PREFACE	iii
CHAPTER ONE	**FUNDAMENTAL DEFINITIONS AND CONCEPTS**	**1**
1.0	Mass Transfer	1
1.1	Classes of Mass Transfer Operations	2
1.2	Mass Transfer as a Diffusion Process	5
1.3	Mass Transfer as a Rate and Transport Process	12
1.4	Mass Transfer in Flow Systems	25
	REFERENCES	39
CHAPTER TWO	**THEORIES OF MASS TRANSFER AT AN INTERFACE**	**41**
2.1	Formulating Mass Transfer Theories	41
2.2	Basic Definitions and Concepts	41
2.3	Solutions of the Diffusion Equation and the Associated Mass Transfer Theories	44
2.4	Theories of Mass Transfer Incorporating Eddy Diffusion	59
2.5	WORKED EXAMPLE	63
	REFERENCES	66

CHAPTER THREE	INTERFACIAL PROPERTIES AND PHENOMENA	67
3.1	Primary and Secondary Properties and Processes at a Mass Transfer Interface	67
3.2	Interfacial Properties of Pure Liquids	69
3.3	Interfacial Phenomena	80
	REFERENCES	108
CHAPTER FOUR	MASS TRANSFER WITHOUT CHEMICAL REACTION	**109**
4.0	Introduction	109
4.1	Mass Transfer of Component A in Two Phases	109
4.2	Sizing Mass Transfer Process Equipment	115
4.3	WORKED EXAMPLES	134
	REFERENCES	146
CHAPTER FIVE	MASS TRANSFER WITH CHEMICAL REACTION	**147**
5.0	Introduction	147
5.1	Conservation of Mass in Diffusion with Chemical Reaction in Homogenous Medium	147
5.2	Chemical Kinetics	149
5.3	The General Case of Diffusion with Chemical Reaction on the Basis of the Whitman Film Theory.	150
5.4	Unsteady State Diffusion with Chemical Reaction on the Basis of the Penetration Theory.	181
5.5	Mass Transfer with Chemical Reaction in Heterogeneous Medium	193

	REFERENCES	210
APPENDIX I	THE GENERAL DIFFUSION EQUATION	211
APPENDIX II	THE ERROR FUNCTION	215
APPENDIX III	THE LAPLACE TRANSFORM	219
APPENDIX IV	SURFACE TENSION	231
APPENDIX V	VISCOSITY TABLES	239

CHAPTER ONE
FUNDAMENTAL DEFINITIONS AND CONCEPTS

1.0: Mass Transfer

Mass transfer (Wikipedia, 2008) is the phrase commonly used in engineering for physical processes that involve molecular and convective transport of atoms and molecules within physical systems. Mass transfer includes both fluid flow and separation unit operations.

Some common examples of mass transfer processes are the evaporation of water from a pond to the atmosphere; the diffusion of chemical impurities in lakes, rivers, and oceans from natural or artificial point sources; mass transfer is also responsible for the separation of components in an apparatus such as a distillation column. In heating, ventilation and air conditioning (HVAC) examples of a heat and mass exchangers are cooling towers and evaporative coolers where evaporation of water cools that portion which remains as a liquid, as well as cooling and humidifying the air passing through.

The driving force for mass transfer is a difference in concentration; the random motion of molecules causes a net transfer of mass from an area of high concentration to an area of low concentration. The amount of mass transfer can be quantified through the calculation and application of mass transfer coefficients. Mass transfer finds extensive application in chemical engineering problems, where material balance on components is performed.

In astronomy, mass transfer is the process by which matter, gravitationally bound to a body, usually a star, fills its Roche lobe and becomes gravitationally bound to a second body, usually a compact object (white dwarf, neutron star or black hole), and is eventually accreted onto it. It is a common phenomenon in binary systems, and may play an important role in some types of supernovae, and pulsars.

For separation processes, thermodynamics determines the extent of separation, while mass transfer determines the rate at which the separation will occur.

1.1: Classes of Mass Transfer Operations

Many classifications of mass transfer processes and operations are possible. The one that is used at any particular time will depend, essentially, on how convenient it is and how applicable it is to the problem at hand. Thus, a classification can be based on the phenomenon or mechanism of the particular mass transfer, on the type of materials participating in the mass transfer process or on the industrial operation that needs to be performed to carry out the mass transfer. It can, also, be classified as a rate process or on whether chemical reaction, also, takes place during, or driving, the mass transfer. It should be noted, however, that no particular classification, necessarily, excludes another.

1.1.1: Classifications Based on Mass Transfer Mechanism or Phenomenon

While classical thermodynamic concepts such as fugacity and chemical potential could be used to enrich classifications based on phenomena, we shall restrict ourselves to their consequences such as molecular diffusion and transport. Thus when mass transfer phenomenon is considered a **Diffusion Process** we are concerned that, in laminar flow and stagnant fluids, mass transfer occurs mostly by *molecular diffusion*, occasioned by Brownian motion while in turbulent flow, it occurs by *eddy diffusion*. When mass transfer is considered as a **Transport Process** we can trace the similarity of the transport of mass, in mass transfer, to the transport of heat, in heat transfer, and to the transport of momentum in momentum transfer, even though they may have occurred as a result of molecular diffusion. Again, considerations of laminar and turbulent flow can be made together with the use of the concepts of molecular and eddy diffusivity.

CHAPTER ONE: FUNDAMENTAL DEFINITIONS AND CONCEPTS

1.1.2: Classifications Based on the Nature of the Contacting Phases

There are three main classes of mass transfer operations when consideration of the contacting phases is of primary importance.

Contact of Immiscible Phases

Here mass transfer occurs across an interface separating the phases. Common examples of this kind of contacting include the transfer of mass between
i. a gas and a liquid phase, as happens in distillation, humidification, dehumidification, gas absorption or desorption etc.
ii. a gas and a solid phase as happens in adsorption, fractional sublimation, etc.
iii. a liquid and a liquid phase such as in solvent extraction
iv. a liquid and a solid phase such as in leaching.

Gas/gas operations are not used, even though they fall within this classification because such solutions have infinite solubility rates. Similarly, solid/solid solutions, at normal temperatures, are of very little practical use because of extremely slow transfer rates.

Contact of Miscible Phases Separated by a Semi-permeable Membrane

Here mass transfer occurs between miscible phases through a semi-permeable membrane which separates them. Examples of this kind of transfer are
i. gas/gas transfers, such as occur in gaseous diffusion and effusion
ii. liquid/liquid transfers, such as occur in *Dialysis*, where a crystalline solid is separated from a colloid by contact of its solution with a membrane permeable only to the crystalline substance, in *Electro-dialysis* where the above process is assisted by an electric potential difference, in *Osmosis* in which a solution is separated from a pure solvent by a membrane,

permeable only to the solvent and in *Electro-osmosis*, where the above process is assisted by an electric potential difference.

iii   Again, there is no solid/solid operation where the solids are separated by a semi-permeable membrane for the purpose of effecting a transfer of material between them.

Direct Contact of Miscible Phases

This type of contacting is used, in commercial practice, for small scale or batch mixing operations, sometimes, preparatory to more difficult mass transfer operations. It is not often necessary to do more than determine the mixing time (to reach a desired consistency) because of the fact that the concentration gradient, necessary for mass transfer, is quickly lost as soon as the phases are mixed. Exceptions arise, however, whenever another type of gradient to drive the mass transfer can be used, such as a temperature gradient, by convection, as in ***Thermal Diffusion*** used to separate uranium isotopes.

1.1.3: Classifications Based on Operational Methods

The major classifications based on operational methods are:

Unsteady State Operations

Here, the transfer of mass is dependent on time of contact. Some industrial processes take place in a ***Batch*** or ***Semi-batch*** manner.

Steady State Operations

Here, the mass transfer rate is independent of time. Common steady state industrial processing modes are the ***co-current*** mode when the streams, exchanging mass with each other, flow parallel to each other in the same direction, *the **counter-current**,* when they flow parallel but in opposite direction to each other and ***cross-flow*** when they flow at right angles or near right angles to each other.

CHAPTER ONE: FUNDAMENTAL DEFINITIONS AND CONCEPTS

Stage-wise Operations

Here, the phases are contacted and separated after mass transfer has occurred between them, within a unit called a *stage*. Such stages are also regarded as *ideal* if the, theoretically, maximum possible mass transfer occurred in them and *equilibrium* stages if some sort of maximum(ideal or practical) possible transfer is established in equilibrium. They may be likened to several batch operations in series.

Continuous or Differential Contact

In this mode, the phases are continuously contacted with each other, within the same unit, so that mass transfer occurs differentially between them as the contacting takes place. It is, often, a steady state operation. Co-current and counter-current contacting can, also, take place in this mode.

1.1.4: Classifications When Chemical Reactions Occur

When chemical reactions occur, analysis becomes more complex since consideration of the order and mechanisms of the reaction, as well as interactions between reaction and diffusion processes, have to be made in addition. Treatment of mass transfer with and without chemical reaction will be made in separate but subsequent chapters.

1.2: Mass Transfer as a Diffusion Process

Diffusion, as a process, is the random motion of matter, in small particle form, from a region of high concentration to one of a lower concentration. Molecular diffusion is said to be due to Brownian motion and predominates in stagnant media or in fluids in laminar flow. Eddy diffusion arises from turbulence, a consequence of some physical property instability, and predominates in turbulent flow or agitation of fluids.

Mathematical analysis of diffusion is best begun with a consideration of the diffusion in a system consisting only of two species of matter which we shall designate as A and B. Let us, also, base our analysis on specie A. According to Maxwell and Stephan, the change in concentration of specie A, dC_A, at any point or time in the system will depend on three factors, namely, the relative velocity with which the units of A in the system move with respect to those of B, $(u_A - u_B)$, the molecular concentrations of A and B in the system C_A, C_B, and the distance through which diffusion of A occurs, dz. Expressed mathematically, this gives:

$$-dC_A = \beta C_A C_B (u_A - u_B) dz \qquad (1.01)$$

where, β is a constant.

1.2.1: Molecular Diffusion in Gases

For gases, specifically, an ideal gas;

$$PV = nRT \qquad (1.02)$$

Where P is the pressure of the gas, V its volume, n the number of moles of gas present, R the gas constant and T its absolute temperature. From this can be derived

$$\frac{P}{RT} = \frac{n}{V} = C = \frac{\rho}{M} \qquad (1.03)$$

where C is now the molar concentration, ρ the molar density and M its molecular weight (formula weight). Using equation (1.03) in equation (1.01), we get

$$-dP_A = \beta \frac{\rho_A}{M_A} \frac{\rho_B}{M_B} (u_A - u_B) dz \qquad (1.04)$$

$$-dP_A = \beta \left[\frac{\rho_A u_A}{M_A} \frac{\rho_B}{M_B} - \frac{\rho_B u_B}{M_B} \frac{\rho_A}{M_A} \right] dz \qquad (1.05)$$

By defining molar flux as, N,

$$N = \text{molar flux} = \text{moles diffusing}/s.m^2$$

$$= \frac{\rho u}{M} \qquad (1.06)$$

CHAPTER ONE: FUNDAMENTAL DEFINITIONS AND CONCEPTS

$$-dP_A = \beta\left[N_A \frac{\rho_B}{M_B} - N_B \frac{\rho_A}{M_A}\right] \quad (1.07)$$

From equations (1.03) and (1.07)

$$-dP_A = \frac{\beta}{RT}(N_A P_B - N_B P_A)dz \quad (1.08)$$

From equations (1.03) and (1.08), knowing that the total pressure, P_T is the sum of the partial pressures, P_A and P_B, of components A and B, that is

$$P_T = P_A + P_B \quad \text{and} \quad P_B = P_T - P_A \quad (1.09)$$

equation (1.08) becomes

$$-dP_A = \frac{\beta}{RT}(N_A P_T - N_A P_A - N_B P_A)dz$$

$$= \frac{\beta}{RT}(N_A P_T - (N_A + N_B)P_A)dz \quad (1.10)$$

For a gas, the binary diffusion coefficient is given by

$$D_{AB} = \frac{RT}{\beta P_T} \quad (1.11)$$

Thus

$$-dP_A = \frac{RT}{D_{AB} P_T}(N_A P_T - (N_A + N_B)P_A)dz \quad (1.12)$$

1.2.1.1: Equimolar Counter Diffusion (EMCD)

One particular system of interest is that in which components, A and B, in the system, diffuse, at steady state, at the same molar flux rate but in opposite directions. Such a situation arises in processes like distillation. The molecules of A and B are said to diffuse in equimolar counter diffusion. That is

$$N_A = -N_B = constant \quad (1.13)$$

When we substitute equation (1.13) into equation (1.12) we get that

$$-dP_A = \frac{RT}{D_{AB} P_T}(N_A P_T)dz = \frac{RT}{D_{AB}}.N_A dz$$

Hence

$$-\int_{P_{A_1}}^{P_{A_2}} dP_A = \frac{RT N_{Az}}{D_{AB}} \int_0^z dz \quad \text{or} \quad N_{Az} = \frac{D_{AB}}{RT z}(P_{A_1} - P_{A_2}) \quad (1.14)$$

Equation (1.14) can be illustrated, graphically, to show how the partial pressures of the two components, A and B, vary over the thickness, z, through which each is diffusing at equal rates but in opposite direction. This is shown in Figure 1.1 below.

Figure 1.1:   Concentration Profiles in Equimolar Counter Diffusion

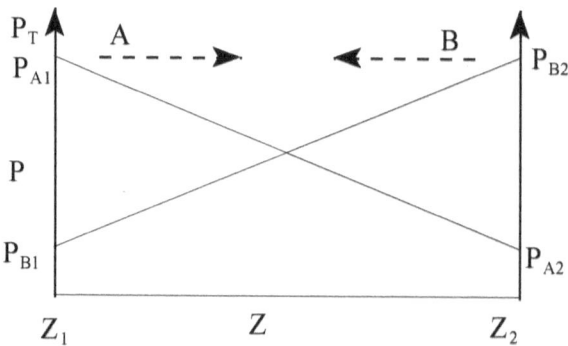

1.2.1.2: Steady State Diffusion Through A Stagnant, Non – Diffusing Medium

In this case, only one component, say A, is diffusing while the other component, B, is stagnant. Such a case arises in humidification and dehumidification where water vapour is the diffusing component whilst air is regarded as being non – diffusing or stagnant. In such a case,

$$N_A = \text{constant}; \quad N_B = 0 \quad (1.15)$$

Substituting equation (1.15) into equation (1.12), we get

$$-dP_A = \frac{RT}{D_{AB} P_T}(N_A P_T - N_A P_A) dz = \frac{RT N_A}{D_{AB} P_T} \cdot (P_T - P_A) dz$$

which, on integration, gives

$$\therefore -\int_{P_{A_1}}^{P_{A_2}} \frac{dP_A}{P_T - P_A} = \frac{RT N_{Az}}{D_{AB} P_T} \int_0^z dz$$

or
$$N_{A_z} = \frac{D_{AB}P_T}{RTz}\ln\left(\frac{P_T - P_{A_2}}{P_T - P_{A_{11}}}\right) \quad (1.16)$$

Since $P_{B_2} = P_T - P_{A_2}$ and $P_{B_1} = P_T - P_{A_1}$, (1.17)

$$P_{B_2} - P_{B_1} = P_{A_1} - P_{A_2} \quad (1.18)$$

From equations (1.17), (1.18) and (1.16)

$$N_A = \frac{D_{AB}P_T}{RTz}\left(\ln\frac{P_{B_2}}{P_{B_1}}\right)\left(\frac{P_{A_1} - P_{A_2}}{P_{B_2} - P_{B_1}}\right)$$

$$= \frac{D_{AB}P_T}{RTzP_{BM}}(P_{A_1} - P_{A_2}) \quad (1.19)$$

where $P_{BM} = \dfrac{P_{B_2} - P_{B_1}}{\ln\dfrac{P_{B_2}}{P_{B_1}}}$ (1.20)

P_{BM} is the log-mean partial pressure difference of B. Equation (1.19) can, also, be stated as

$$N_A = \frac{D_{AB}}{RTz}(P_{A_1} - P_{A_2})\phi \quad (1.21)$$

where $\phi = \dfrac{P_T}{P_{BM}}$ is called the DRIFT FACTOR (1.22)

The drift factor can, also, be seen as the factor by which equimolar counter diffusional flux is multiplied to give the actual flux for diffusion in a stagnant and non – diffusing medium. The concentration profiles for the two components, in a binary system, are illustrated in Figure 1.2 below.

Figure 1.2: Concentration Profiles for Diffusion in a Stagnant Medium

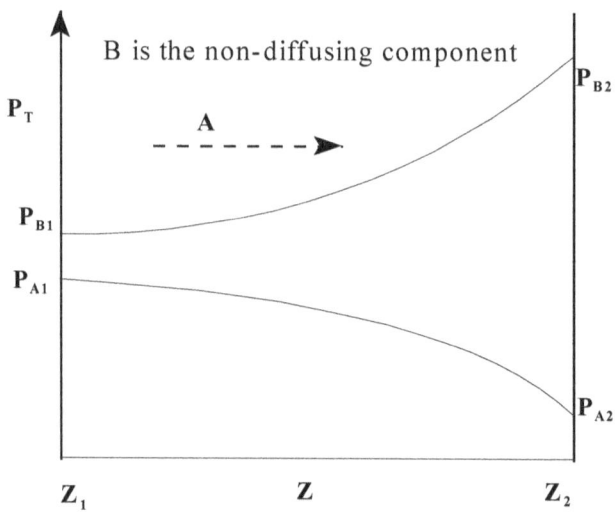

1.2.1.3: Steady State Diffusion Through A Stagnant Multi-Component Gas Mixture

In this case, equation (1.20) still applies but a mean D_{AM}, to account for the contribution of each component of the gas mixture, is used instead of the binary D_{AB}. This mean D_{AM} is obtained, from Physical Chemistry principles, based on component A, as

$$D_{AM} = \frac{1}{\dfrac{Y_B}{D_{AB}} + \dfrac{Y_C}{D_{AC}} + \dfrac{Y_D}{D_{AD}} + \ldots} \qquad (1.23)$$

where Y is the mole fraction of each component on an A-free basis.

1.2.2: Molecular Diffusion in Liquids

Molecular diffusion in liquids follows, also, the Maxwell – Stefan relation (equation (1.01))

$$-dC_A = \beta C_A C_B (u_A - u_B) dz \qquad (1.01)$$

CHAPTER ONE: FUNDAMENTAL DEFINITIONS AND CONCEPTS

The molar flux, N, is
$$N = Cu \quad (1.24)$$
where C is the concentration, moles per meter^3, and u is the molecular velocity, meters per second. Using equation (1.24) in (1.01), we get;
$$-dC_A = \beta(C_B N_A - C_A N_B)dz \quad (1.25)$$

1.2.2.1: Steady State Equimolar Counter-Diffusion

Since N_A and N_B are constant and $N_A = -N_B$, equation (1.25) becomes
$$-dC_A = \beta(C_A + C_B)N_A\,dz \quad (1.26)$$
Since the total concentration, C_T, at any time, is the sum of C_A and C_B,
$$C_T = C_A + C_B \quad (1.27)$$

Define $$D_{AB} = \frac{1}{\beta C_T} \quad (1.28)$$

Then
$$\therefore -dC_A = \frac{N_A}{D_{AB}}dz \quad \text{or} \quad -\int_{C_{A_1}}^{C_{A_2}} dC_A = \frac{N_{A_z}}{D_{AB}}\int_0^z dz$$

i.e. $$N_{A_z} = \frac{D_{AB}}{z}(C_{A_1} - C_{A_2}) \quad (1.29)$$

Equation (1.29) can, also, be expressed in terms of mole fractions
$$N_{A_z} = \frac{D_{AB} C_T}{z}(x_{A_1} - x_{A_2}) \quad (1.30)$$
where $$C_A = x_A C_T \quad (1.31)$$

1.2.2.2: Steady State Diffusion of Component A through a Stagnant Medium

Here N_A is constant while $N_B = 0$.
Equation (1.26) becomes, using equations (1.27) and (1.28),
$$-dC_A = \frac{N_A}{D_{AB} C_T} C_B\,dz \quad (1.32)$$

Since $C_B = C_T - C_A$, equation (1.32) becomes, on integration,

$$\int_{C_{A_1}}^{C_{A_2}} \frac{dC_A}{C_T - C_A} = \frac{N_{A_z}}{D_{AB} C_T} \int_0^z dz$$

or $$N_{A_z} = \frac{D_{AB} C_T}{z} \ln\left(\frac{C_T - C_{A_2}}{C_T - C_{A_1}}\right) \quad (1.33)$$

Equation (1.33) can be rearranged to give

$$N_{A_z} = \frac{D_{AB}}{z}(C_{A_1} - C_{A_2})\frac{C_T}{C_{BM}}$$

$$= \frac{D_{AB}}{z}(x_{A_1} - x_{A_2})\frac{C_T}{x_{BM}} \quad (1.34)$$

1.3: Mass Transfer as a Rate and Transport Process

Mass transfer is a transport process. It is, also, a rate process. A rate process can be expressed, generally, as

$$\text{Rate} = \frac{\text{Driving Force as Potential Difference}}{\text{Resistance to Flow}} \quad (1.35)$$

$$= \text{Driving Force} \times \text{Conductance} \quad (1.36)$$

In comparison with electrical circuit theories,

$$\text{Conductance} = \frac{1}{\text{Resistance}} \quad (1.37)$$

$$= \text{Conductivity} \times \frac{\text{Cross Sectional Area}}{\text{Length of Path Traversed}} \quad (1.38)$$

In a transport process, flux is defined as rate per unit area. Thus

$$\text{Flux} = \frac{\text{Rate}}{\text{Cross Sectional Area}} \quad (1.39)$$

$$= \frac{\text{Driving Force} \times \text{Conductivity} \times \text{Cross-sectional Area}}{\text{Cross-sectional Area} \times \text{Length of Path Traversed}}$$

$$= \frac{\text{Driving Force} \times \text{Conductivity}}{\text{Length of Path Traversed}} \quad (1.40)$$

With these basic definitions, we can, now, examine mass flux, heat flux and momentum flux to show that each of these is a transport process.

CHAPTER ONE: FUNDAMENTAL DEFINITIONS AND CONCEPTS

1.3.1: Analogies Between Heat, Mass, And Momentum Transfer (Wikipedia, 2008)

Molecular transport of heat, or mass have many similarities. The molecular diffusion equations, of Newton for momentum, Fourier's for heat, and Fick's for mass, are very similar. There are, therefore, many analogies among these three molecular transport process. A great deal of effort has been devoted in the literature to developing analogies among these three transport processes for turbulent transfer so as to allow prediction of one from any of the others. Reynolds analogy assumes that the turbulent diffusivities are all equal and that the molecular diffusivities μ, ρ and D_{AB} are negligible compared to the turbulent diffusivities. When liquids are present and/or drag is present the analogy is not valid. Other analogies, such as von Karman's and Prandtl's, usually results in poor relations. The most successful and most widely used analogy is the Chilton and Colburn J-factor analogy. This analogy is based on experimental data for gases and liquids in both the laminar and turbulent regions. Although it is based on experimental data, it can be shown to satisfy the exact solution derived from laminar flow over a flat plate.

1.3.1.1: Mass Transfer by Molecular Diffusion

Fick's law for mass transfer states that the molar flux, n_A, by molecular diffusion, through a medium of thickness, z, is from a region of high to one of lower concentration (C_A) and is, directly, proportional to the concentration gradient. That is;

$$n_{A_z} = -D_{AB} \frac{dC_A}{dz} \qquad (1.41)$$

It can be seen that, since dC_A is a driving force, dz a length of path traversed and D_{AB} a diffusivity (or a conductivity), mass transfer is both a rate and a transport process. Because mass transfer is, often, associated with heat and momentum transfer, it will, also, be shown that these are, also, rate and transport processes.

1.3.1.2: Heat Transfer by Molecular Conduction

Fourier's first law of heat conduction states that the rate, q_z, at which heat energy crosses an isothermal surface, in conduction, and in the direction of falling temperature is directly proportional to the temperature gradient. That is

$$q_z = -k_z \frac{dT}{dz} \qquad (1.42)$$

Where constant of proportionality, k_z, is the thermal conductivity of the material in the direction of heat energy flow. Equation (1.42) can be rewritten as

$$q_z = -\frac{k_z}{\rho C p} \frac{d\rho C p T}{dz} = -\alpha \frac{dH}{dz} \qquad (1.43)$$

Where ρ, Cp, α and H are the density, specific heat capacity at constant pressure, thermal diffusivity and thermal capacity, respectively, of the material. It can be seen, also, that heat transfer is both a rate and a transport process with α as the thermal diffusivity. Similarly, we can analyse momentum transfer in fluid flow.

1.3.1.3: Momentum Transfer in Fluid Flow

Newton's law, for the flow of fluids, states, that for Newtonian fluids, the shear stress in a fluid is, directly, proportional to the shear strain. That is

$$\tau_z = -\mu_z \frac{du}{dz} \qquad (1.44)$$

where τ_z is the shear stress, μ_z is the fluid viscosity and u the fluid velocity. Knowing that shear stress is force per unit area, that force is the rate of change of momentum and that rate of strain is the rate of change of velocity with distance, equation (1.44) can be rewritten as an equation of momentum flux, thus;

$$\tau_{zy} = -\frac{\mu}{\rho} \frac{d(\rho u)}{dz} = -v \frac{dM}{dz} \qquad (1.45)$$

where $M = \rho u$, momentum per unit area and $v = $ momentum diffusivity, also known as the kinematic viscosity. Thus momentum transfer can be seen, from equation (1.45), to be, also, both a rate and a transport process.

1.3.2: Dimensionless Groups in Mass Transfer

Considerations of mass, heat and momentum transfer as transport processes give impetus to the notion that they must be related in some manner in actual physical situations. This relationship is easily illustrated by considering dimensionless groups, which provide indices of the relative effects or importance of one transport mechanism over another in transport phenomena. Table 1.1 lists some of the most important dimensionless groups in mass, heat and momentum transfer.

1.3.3: The Molecular Diffusion Coefficient

Let us look more closely at what happens when mass is being transferred by molecular diffusion in a system in order to gain a better understanding of the concept of molecular diffusivity. Consider two components A and B partitioned in an enclosure as shown below at a total pressure P_T.

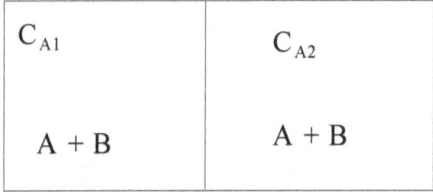

Let. the concentration of A in the first compartment, C_{A1}, be greater than that in the second compartment C_{A2}. Assume, also, that C_A is proportional to the partial pressure of A in each compartment. If the partition is removed, A and B will move as shown below.

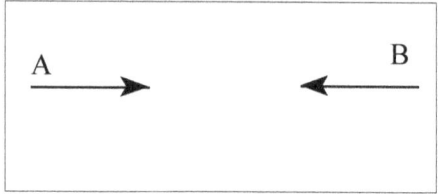

As the total pressure is constant, movement is due only to molecular diffusion and each molecular movement of A means one molecular movement of B. Since P_T is constant

$$n_{AZ} + n_{BZ} = 0 \qquad (1.46)$$

because
$$n_{AZ} = -n_{BZ} \qquad (1.47)$$

From Fick's first law, equation (1.41), and equation (1.47)

$$-D_{AB}\frac{dC_A}{dz} = -\left(-D_{BA}\frac{dC_B}{dz}\right) \qquad (1.48)$$

But $P_T = P_A + P_B$ = constant and $C_A + C_B = C_T$ [equations (1.09) and (1.27)], respectively, and since

$$C_A = \frac{moles\ of\ A}{V} = \frac{n_A}{V} = \frac{P}{RT} \qquad (1.03)$$

From equation (1.27)

$$\frac{dC_A}{dz} = -\frac{dC_B}{dz} \qquad (1.49)$$

When equation (1.49) is substituted in (1.48), it is found that

$$D_{AB} = D_{BA} = D \qquad (1.50)$$

1.3.3.1: Molecular Diffusivity in the Gas Phase

The kinetic theory of gases predicts that, in an ideal gas, the molecular diffusivity is related to the mean molecular velocity and the mean free path for gaseous molecule collisions as:

$$D = \frac{1}{3}\overline{u_m}\overline{\lambda} \qquad (1.51)$$

where $\overline{u_m}$ is the mean molecular velocity and $\overline{\lambda}$ is the mean free path for gaseous molecule collisions. The mean molecular velocity is not easy to measure directly but is related, using the

CHAPTER ONE: FUNDAMENTAL DEFINITIONS AND CONCEPTS

assumptions of the same kinetic theory, to the mean square velocity Since the kinetic theory predicts that

$$\overline{u}^2 = \frac{8}{3\pi}\overline{u_m^2} = \frac{8}{3\pi}\cdot\frac{3RT}{M} = \frac{8RT}{\pi M} \qquad (1.52)$$

where

$$\overline{u_m^2} = \text{the mean square velocity } \frac{3RT}{M} = \frac{3kT}{m} \qquad (1.53)$$

and k is the Boltzman constant (=R/N). R is the universal gas constant, M, the molecular mass, m, the mass of a single molecule and N, the Avogadro's number.

Thus, from equation (1.52)

$$\overline{u_m} = 2\sqrt{\frac{2RT}{\pi M}} \qquad (1.54)$$

Table 1.1a: Some Important Dimensionless Groups in Mass Transfer

Group Name	MASS TRANSFER Physical Significance	Representative Symbol	Mathematical Form
Schmidt Number	Momentum Diffusivity / Mass Diffusivity	N_{Sc} or Sc	Sc = ν/D
Lewis Number	Thermal Diffusivity / Mass Diffusivity	N_{Le} or Le	Le = α/D
Peclet Number (for mass transfer)	Inertia Forces x Momentum Diffusivity / Viscous Forces x Mass Diffusivity	N_{Pe} or Pe	Pe_D = Re. Sc
Sherwood Number	Bulk Convective Mass Transfer / Diffusional Mass transfer	N_{Sh} or Sh	Sh = $\underline{Fd}$ ($C_T D_{AB}$)
Stanton Number (for mass transfer)	Boundary Layer Mass Transfer / Bulk Fluid Mass Transfer	N_{St} or St	St_D=Sh/(Re.Sc) = Sh/ Pe_D
J-factor (for mass transfer)		j_D	= St_D . $Sc^{2/3}$

Table 1.1b: Some Important Dimensionless Groups in Heat Transfer

HEAT TRANSFER

Group Name	Physical Significance	Representative Symbol	Mathematical Form
Prandtl Number	Momentum Diffusivity / Thermal Diffusivity	N_{Pr} or Pr	$Pr = C_p\mu/k$
Stanton Number (for heat transfer)	Boundary Layer Heat Transfer / Bulk Fluid Heat Convection	N_{St} or St	$St_H = Nu/(Re \cdot Pr)$ $= h/\rho u C_p$
Nusselt Number	Boundary Layer Heat Transfer / Boundary Layer Heat Conduction	N_{Nu} or Nu	$Nu = \dfrac{hd}{k}$
Peclet Number (for heat transfer)	Inertia Forces x Momentum Diffusivity / Viscous Forces x Thermal Diffusivity	N_{Pe} or Pe	$Pe_H = Re \cdot Pr$
J-factor (for heat transfer)		J_H	$St_H \cdot Pr^{2/3}$

Table 1.1c: Some Important Dimensionless Groups in Momentum Transfer

MOMENTUM TRANSFER

Group Name	Physical Significance	Representative Symbol	Mathematical Form
Reynolds Number	Inertia Forces / Viscous Forces	N_{Re} or Re	$Re = \rho u d/\mu$
Froude Number	Inertia Forces / Gravitational Forces	N_{Fr} or Fr	$Fr = u^2/gd$
Weber Number	Inertia Forces / Surface Tension Forces	N_{We} or We	$We = \rho d u^2/(g\sigma)$
Euler Number	Pressure Forces / Inertia Forces	N_{Eu} or Eu	$Eu = g\Delta P/(\rho u^2)$
Friction factor, f		C_f or f	$const._1 \cdot Re^{Const.2}$

Now

$$\lambda = \frac{1}{\sqrt{2}\,\pi n \sigma^2} \qquad (1.55)$$

$$\text{and} \quad P = \frac{1}{3} n m \overline{u^2} = \frac{3kTn}{3} = nkT \quad \text{or} \quad n = \frac{P}{kT} \qquad (1.56)$$

where n is the number of molecules per cm^3 and σ is the collision diameter.

CHAPTER ONE: FUNDAMENTAL DEFINITIONS AND CONCEPTS

From (1.56) and (1.55)

$$\lambda = \frac{kT}{\pi P \sigma^2 \sqrt{2}} \tag{1.57}$$

Substituting (1.57) and (1.54) into (1.51) and re-arranging;

$$D = \frac{1}{3} \cdot 2 \cdot \sqrt{\frac{2RT}{\pi M}} \cdot \frac{kT}{\pi P \sigma^2 \sqrt{2}} = \frac{2}{3N} \left(\frac{R}{\pi}\right)^{1.5} \frac{T^{1.5}}{P\sigma^2} \cdot \sqrt{\frac{1}{M}}$$

$$= cons \tan t. \frac{T^{1.5}}{P\sigma^2} \cdot \sqrt{\frac{1}{M}} \tag{1.58}$$

For self diffusion,

$$D_{AA} = cons \tan t. \frac{T^{1.5}}{P\sigma_A^2} \cdot \sqrt{\frac{1}{M_A}} \tag{1.59}$$

For binary diffusion,

$$D_{AB} = cons \tan t. \frac{T^{1.5}}{P\sigma_{AB}^2} \cdot \sqrt{\frac{1}{M_{AB}}} \tag{1.60}$$

where, in the so called Gilliland's method

$$\sigma = \left(\frac{\sum V}{N}\right)^{\frac{1}{3}} \tag{1.61}$$

$$\sigma_{AB} = \frac{1}{2N^{\frac{1}{3}}}\left[\left(\sum V_A\right)^{\frac{1}{3}} + \left(\sum V_B\right)^{\frac{1}{3}}\right] \tag{1.62}$$

$$\frac{1}{M_{AB}} = \frac{1}{2}\left(\frac{1}{M_A} + \frac{1}{M_B}\right) \tag{1.63}$$

Thus, substituting (1.60) and (1.59) in (1.57), we get

$$D_{AB} = cons \tan t \frac{T^{1.5}\sqrt{\frac{1}{M_A} + \frac{1}{M_B}}}{P\left[\left(\sum V_A\right)^{\frac{1}{3}} + \left(\sum V_B\right)^{\frac{1}{3}}\right]^2} \tag{1.64}$$

Equation (1.64) is symmetrical in A and B hence $D_{AB} = D_{BA}$. If T is in degrees Kelvin, P in atmospheres, V in cm^3/g.mole and D in cm^2/second, then the constant in equation (1.64) is equal to 0.0043. Perry and Green (1984) report a slightly different expression, (equation 1.65) and give values of atomic and molecular volumes, as shown in Table 1.2 below, to be used, for simple molecules.

$$D_{AB} = 0.001 \frac{T^{1.75}\sqrt{\frac{1}{M_A} + \frac{1}{M_B}}}{P\left[\left(\sum V_A\right)^{1/3} + \left(\sum V_B\right)^{1/3}\right]^2} \quad (1.65)$$

Equation (1.65) predicts the diffusion coefficient, for non-polar gases, up to 10 atmospheres pressure, with accuracy of between 5 and 10%. No composition dependence is reported.

For mixtures of non-polar gases or a polar gas with a non-polar gas, Treybal (1980) recommends the Wilke – Lee modification of the Hirschfelder-Bird-Spotz method which results in the equation;

$$D_{AB} = \frac{10^{-4}\left[1.084 - 0.249\sqrt{\frac{1}{M_A} + \frac{1}{M_B}}\right]T^{1.5}\sqrt{\frac{1}{M_A} + \frac{1}{M_B}}}{P_T \sigma_{AB}^2 f\left(\frac{kT}{\varepsilon_{AB}}\right)} \quad (1.66)$$

where D_{AB} is the diffusivity, m²/s; T, the absolute temperature, deg. K; P_T the absolute pressure, N/m²; M_A, M_B, the molecular weights of A and B, respectively, kg/kmol; σ_{AB}, the molecular separation at collision = $(\sigma_A + \sigma_B)/2$, nm; ε_{AB}, the energy of molecular attraction = $\sqrt{(\varepsilon_A.\varepsilon_B)}$; k, the Boltzman's constant and $f(kT/\varepsilon_{AB})$, the collision function, usually given as a curve but listed in Table 1.3, below, at a few points in the range of values for which it is valid.

The molecular separation, σ, at collision, can be estimated from equation (1.64) while the ratio, ε/k, can be estimated from equation (1.65).

$$\sigma = 1.18 V^{1/3} \quad (1.67)$$

$$\frac{\varepsilon}{k} = 1.21 T_B \quad (1.68)$$

V is estimated from atomic or molecular volumes as in Table 1.4.

CHAPTER ONE: FUNDAMENTAL DEFINITIONS AND CONCEPTS

Table 1.2: Atomic and Molecular Diffusion Volumes for Estimating Diffusion Coefficients according to the Method of Fuller, Schettler and Giddings (Perry & Green, 1984)

Atomic Diffusional Volumes		Molecular Diffusional Volumes	
Element	Atomic Volume	Molecule	Molecular Volume
C	16.5	H_2	7.07
H	1.98	N_2	17.9
O	5.48	O_2	16.6
N	5.69	Air	20.1
Cl	19.5	CO	18.9
S	17.0	CO_2	26.9
Aromatic ring	-20.2	N_2O	35.9
Hetero-cyclic ring	-20.2	NH_3	14.9
		H_2O	12.7
		SO_2	41.1

Table 1.3: Values of the Collision Function $f(kT/\varepsilon_{AB})$ (Treybal, 1980)

kT/ε_{AB}	$f(kT/\varepsilon_{AB})$	kT/ε_{AB}	$f(kT/\varepsilon_{AB})$
0.3	1.34	6.0	0.41
0.4	1.16	10	0.37
0.5	1.04	20	0.335
0.6	0.94	40	0.30
1.0	0.72	60	0.28
1.5	0.60	100	0.26
2.0	0.54	200	0.232
4.0	0.44	400	0.21

Thus, for Benzene, C_6H_6, V is estimated to be

$$6(0.0148) + 6(0.0037) - 0.015 = 0.096.$$

This gives

$$\sigma_{Benzene} = 1.18 \times (0.096)^{1/3} = 0.4579 \; nm.$$

Table 1.4: Atomic and Molecular Volumes for Estimating the Diffusion Coefficient of Gases (Treybal, 1980)

Element/Compound	Atomic Volume, m³/1000 atoms x 10³	Molecular Volume, m³/kmol x 10³	Element/Compound	Atomic Volume, m³/1000 atoms x 10³	Molecular Volume, m³/kmol x 10³
Carbon	14.8		Benzene ring	-15	
Hydrogen	3.7	14.3	Naphthalene ring	-30	
Chlorine	24.6	48.4	H_2S		32.9
Sulphur	25.6				
Nitrogen	15.6	31.2			
In primary amines	10.5		Air		29.9
In secondary amines	12.0		CO		30.7
Oxygen	7.4	25.6	CO_2		34.0
In methyl ethers	9.1		N_2O		36.4
In higher esters	11.0		NH_3		25.8
In acids	12.0		H_2O		18.9
In methyl ethers	9.9		SO_2		44.8
In higher ethers	11.0		NO		23.6

In practice, experimentally determined values of molecular diffusivity are preferred for engineering use. Typical experimental values, for diffusion in air, are shown in Table 1.5.

Table 1.5: Experimental Values of the Diffusion Coefficient of Gases in Air at Atmospheric Pressure (Bolz & Tuve, 1973), (Treybal, 1980)

Gas	Diffusion Coefficient at 0 C, m²/s x 10⁵	Diffusion Coefficient at 25 C, m²/s x 10⁵	Diffusion Coefficient at 59 C, m²/s x 10⁵
H_2	6.11	7.12	

N₂	1.78		
O₂	1.78	2.06	
CO₂	1.42	1.64	
NH₃	1.98	2.29	
H₂O	2.18	2.53	3.05
Aniline	0.61	0.72	0.90
Benzene	0.751	0.88	
n-Butanol	0.703	0.90	1.04
Chlorobenzene		0.73	0.90
Ethanol	1.02	1.19	
Ethyl Acetate	0.715	0.85	1.06
Toluene	0.709	0.84	0.92

1.3.3.2: Molecular Diffusivity in the Liquid Phase

The basis of the theory, in this case, is the equation for the hydrodynamics of a sphere of component, A, moving in laminar flow but with terminal velocity, U_A, through a fluid, B, of viscosity, μ_B, solely as a result of molecular diffusion. The force due to diffusional pressure or concentration gradient is given by

$$F = \frac{kT}{D_{AB}} U_A \qquad (1.69)$$

The frictional force, resisting this movement, F_R, is given, from hydrodynamic considerations, by

$$F_R = 3\pi\mu_B\sigma_A U_A \frac{2\mu_A + \beta_{AB}\frac{\sigma_A}{2}}{3\mu_B + \beta_{AB}\frac{\sigma_A}{2}} \qquad (1.70)$$

where β_{AB} is the coefficient of sliding friction, μ_B, the viscosity of the medium, σ_A, the equivalent of a collision distance term used in the case of collision of gas molecules and U_A, the velocity of particle motion. When no slip occurs, $\beta_{AB} = \infty$ and

$$F_R = 3\pi\mu_B\sigma_A U_A \qquad (1.71)$$

Equation (1.71) is seen to be Stoke's law for the motion of a single particle through a fluid. Under steady state conditions, $F = F_R$ so that, from equations (1.69), (1.71) and (1.61),

$$\frac{D_{AB}\mu_B}{T} = \frac{k}{3\pi\sigma_A} \approx \frac{k}{3\pi}\left(\frac{N}{V}\right)^{1/3} = \text{constant} \qquad (1.72)$$

Equation (1.72) is found to be valid for situations in which the diffusing molecules are large compared to the molecules of the medium in which they are diffusing. When there is slip between diffusing molecules and those of the medium, $\beta_{AB} = 0$ and

$$F_R = 2\pi\mu_B\sigma_A U_A \qquad (1.73)$$

Under steady state conditions, $F = F_R$ so that, from equations (1.69), (1.73) and (1.61),

$$\frac{D_{AB}\mu_B}{T} = \frac{k}{2\pi\sigma_A} \approx \frac{k}{2\pi}\left(\frac{N}{V}\right)^{1/3} = \text{constant} \qquad (1.74)$$

Equation (1.74) is valid for self-diffusion. Experimentally determined values of diffusivities (see Table 1.6) are always preferable to those estimated from theoretical considerations. When these are not available, however, Wilke and Chang (Treybal, 1980) recommend the empirical expression which predicts diffusivities in dilute solutions, usually, aqueous non-electrolytes, with accuracies of between 6.9 to 10%.

$$D_{AB} = \frac{117.3 \times 10^{-18}(\phi M_B)^{0.5} T}{\mu V_A^{0.6}} \qquad (1.75)$$

where μ is the viscosity of the solution, kg/m.s; M_B, the molecular weight of the solvent, kg, V_A, the molal volume of the solute at boiling point, m^3/kmol = 0.0756 for water as solute; and φ, the association factor = 2.26 for water as solvent, 1.9 for methanol

CHAPTER ONE: FUNDAMENTAL DEFINITIONS AND CONCEPTS

as solvent, 1.5 for ethanol as solvent and 1.0 for unassociated solvents such as benzene and ethyl ether.

For dilute solutions of organic solutes diffusing in organic solvents, Perry and Green (1984) recommend the Scheibel equation given by

$$D_{AB} = \frac{KT}{\mu_B V_A^{1/3}} \quad (1.76)$$

where

$$K = 8.2 \times 10^{-8} \left[1 + \left(\frac{3V_B}{V_A}\right)^{2/3}\right] \quad (1.77)$$

The unit of μ is cP while that of V is cm^3/g.mol. When the solvent is benzene and $V_A < 2V_B$, $K = 18.9 \times 10^{-8}$ whilst for other solvents, if $V_A < 2.5V_B$, $K = 17.5 \times 10^{-8}$. For concentrated solutions, Treybal (1980) recommends equation (1.78)

$$D_A \mu = \left(D_{BA}^0 \mu_A\right)^{x_A} \left(D_{AB}^0 \mu_B\right)^{x_B} \left[1 + \frac{d \log \gamma_A}{d \log x_A}\right] \quad (1.78)$$

where D_{AB}^0 is the diffusivity of A in B at infinite dilution, D_{BA}^0, the diffusivity of B in A at infinite dilution, γ, the activity coefficient and x, the mole fraction.

1.4: Mass Transfer in Flow Systems

So far, we have considered only systems in diffusional flux with no bulk motion of material. This enabled us to define some basic concepts more clearly. A more general case is likely to be that in which bulk motion, in addition to molecular diffusion, occurs. In such a case;

$$\text{Total Molar Flux} = N_{A_z} = \text{Molar Flux of Component A}$$
$$\text{due to bulk flow + Molar Flux}$$
$$\text{due to diffusional flux}$$

$$= \frac{V_z C_A}{S} + n_A \quad (1.79)$$

where V_z is the volumetric flow rate, C_A is the molar concentration per unit volume and n_{Az} is the diffusional molar flux. S is the cross sectional area for flow. Since $U_z = V_z/S$, we can rewrite equation (1.79) as

$$N_{A_z} = U_z C_A + n_{A_z} \qquad (1.80)$$

And since by Fick's law,

$$n_{A_z} = -D_{AB}\frac{dC_A}{dz} \qquad \text{from equation (1.41)}$$

$$N_{A_z} = U_z C_A - D\frac{dC_A}{dz} \qquad (1.81)$$

1.4.1: Mass Transfer by Laminar Diffusion in Binary Systems.

Consider diffusion in a binary mixture made up of components A and B. From equation (1.80)

$$N_{A_z} = U_z C_A + n_{A_z} \qquad (1.80)$$
$$N_{B_z} = U_z C_B + n_{B_z} \qquad (1.80)$$
$$N_{A_z} + N_{B_z} = U_z(C_A + C_B) + n_{A_z} + n_{B_z}$$
$$= U_z C_T \qquad (1.81)$$

since, from (1.27) and (1.46),

$$C_A + C_B = C_T \text{ and } n_{AZ} + n_{BZ} = 0$$

Thus, from equation (1.81)

$$U_z = \frac{N_A + N_B}{C_T} \qquad (1.82)$$

Substituting (1.82) into (1.80)

$$N_{A_z} = \frac{N_{AZ} + N_{BZ}}{C_T}C_A + n_{AZ} = (N_{AZ} + N_{BZ})X_A + n_{A_z} \qquad (1.83)$$

where X_A is the mole fraction of A = C_A/C_T \qquad (1.84).

CHAPTER ONE: FUNDAMENTAL DEFINITIONS AND CONCEPTS

Table 1.6: Experimental Values of the Diffusion Coefficient of Solutes in Water at 20 C (Bolz & Tuve, 1973)

Substance	Diffusion Coefficient, $m^2/s, \times 10^9$	Substance	Diffusion Coefficient, $m^2/s, \times 10^9$	Substance	Diffusion Coefficient, $m^2/s, \times 10^9$
H_2	5.13	HCl	2.64	Acetic Acid	0.88
N_2	1.64	HNO_3	2.6	Ethanol	1.00
O_2	1.80	H_2SO_4	1.73	Glycerol	0.72
Cl_2	1.22	NaOH	1.51	Phenol	0.84
CO_2	1.77	NaCl	1.35	Sucrose	0.45
NH_3	1.76	Acetylene	1.56		

1.4.1.1: Special Cases : Equimolar Counter Diffusion

In equimolar counter diffusion, for each molecule of A that diffuses in one direction, one molecule of B will move in the opposite direction. Mathematically, this is the same as

$$N_{A_z} + N_{B_z} = 0 \quad (1.85)$$

So that, from equation (1.83)

$$N_{A_z} = n_{A_z} \quad i.e.\ U_z = 0 \quad (1.86)$$

That is:

$$molar\ flux = diffusional\ flux \quad (1.87)$$

Or

$$N_{A_z} = n_{A_z} = -D\frac{dC_A}{dz} \quad (1.88)$$

At steady state, and for N_{AZ} constant;

$$N_{A_z}\int_{z_1}^{z_2} dz = -D\int_{C_{A_1}}^{C_{A_2}} dC_A \quad (1.89)$$

This gives, on integration and rearrangement, the value of the bulk molar flux as

$$N_{A_z} = D\frac{C_{A_1} - C_{A_2}}{z_2 - z_1} = \frac{D}{L}(C_{A_1} - C_{A_2}) \quad (1.90)$$

where $L = z_2 - z_1$. At any distance, z, since N_{AZ} is constant, equation (1.90) becomes

$$N_{A_z} = D\frac{C_{A_1} - C_A}{z - z_1} \quad (1.91)$$

Dividing equation (1.90) by equation (1.91) and rearranging, we get that the concentration profile, for equimolar counter diffusion during mass transfer in laminar flow is

$$C_A = C_{A_1} - (C_{A_1} - C_{A_2})\frac{z - z_1}{z_2 - z_1} \quad (1.92)$$

This profile is shown in the diagram below.

Fig 1.3: Concentration Profile in Equimolar Counter-Diffusion

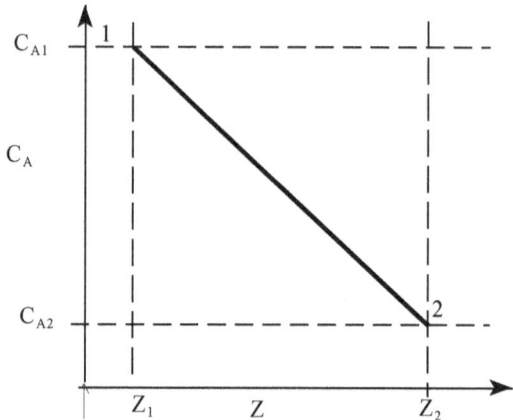

1.4.1.2: Diffusion through a Stagnant Medium

This is the situation in which one component diffuses through a stagnant component in a gas or liquid mixture, through an interface, into another medium. For a gas/liquid system, the situation can be illustrated as shown below.

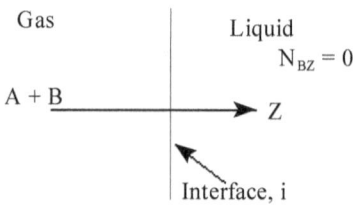

CHAPTER ONE: FUNDAMENTAL DEFINITIONS AND CONCEPTS

If component B is the stagnant component, by definition, there is no molar flux of B. Hence $N_{Bz} = 0$. Note, however, that there is, still, molecular diffusion of B. Equation (1.81) becomes

$$N_{Az} = X_A N_{Az} + n_{Az}$$

from which
$$N_{Az} = \frac{n_{Az}}{1-X_A} = \frac{n_{Az}}{X_B} = -\frac{n_{Bz}}{X_B} \qquad (1.93)$$

Since, from equations (1.88) and (1.84)

$$n_{Az} = -D\frac{dC_A}{dz} \qquad (1.88)$$

and

$$C_A = X_A C_T \qquad (1.84)$$

$$n_{Az} = -DC_T \frac{dX_A}{dz} \qquad (1.94)$$

Similarly,

$$n_{Bz} = -DC_T \frac{dX_B}{dz} \qquad (1.95)$$

From (1.93) and (1.95)

$$N_{Az} = D\frac{C_T}{X_B}\frac{dX_B}{dz} \qquad (1.96)$$

Under steady state conditions, N_{Az} is constant, so that

$$N_{Az}\int_{z_1}^{z_2} dz = DC_T \int_{C_{A_1}}^{C_{A_2}} \frac{dX_B}{X_B} dC_A$$

i.e.
$$N_{Az} = \frac{DC_T}{z_2 - z_1}\ln\frac{X_{B_2}}{X_{B_1}} = \frac{DC_T}{z_2 - z_1}\ln\frac{C_{B_2}}{C_{B_1}} \qquad (1.97)$$

For an ideal gas, $PV = nRT$ and concentration, $C = n/V = P/RT$, so that equation (1.97) becomes

$$N_{Az} = \frac{DP_T}{RT(z_2 - z_1)}\ln\frac{P_{B_2}}{P_{B_1}} \qquad (1.98)$$

In order to reflect the effects of the stagnant component on the molar flux of the diffusing component, a log-mean partial pressure difference, for the non-diffusing gas, P_{BM} is defined as

$$P_{BM} = \frac{P_{B_2} - P_{B_1}}{\ln\frac{P_{B_2}}{P_{B_1}}} \qquad (1.99)$$

It is, easily, seen, from equation (1.99) that

$$\ln \frac{P_{B_2}}{P_{B_1}} = \frac{P_{B_2} - P_{B_1}}{P_{BM}} = \frac{(P_T - P_{A_2}) - (P_T - P_{A_1})}{P_{BM}}$$

$$= \frac{P_{A_1} - P_{A_2}}{P_{BM}} \tag{1.100}$$

Hence, putting (1.100) in (1.98)

$$N_{Az} = \frac{D}{RT(z_2 - z_1)} \frac{P_T}{P_{BM}} (P_{A_1} - P_{A_2}) \tag{1.101}$$

Similarly, since

$$\frac{P_{A_1} - P_{A_2}}{P_{BM}} = \frac{C_{A_1} - C_{A_2}}{C_{BM}} = \frac{X_{A_1} - X_{A_2}}{X_{BM}} \tag{1.102}$$

$$N_{Az} = \frac{D}{(z_2 - z_1)} \frac{C_T}{C_{BM}} (C_{A_1} - C_{A_2})$$

$$= \frac{D}{(z_2 - z_1)} \frac{C_T}{X_{BM}} (X_{A_1} - X_{A_2}) \tag{1.103}$$

The ratio, $C_T/C_{BM} = P_T/P_{BM} = 1/X_{BM} > 1$, is known as the *drift factor* and is equivalent to the ratio of the mass flux during diffusion through a stagnant medium to that in equimolar counter diffusion. The concentration profile during diffusion through a stagnant medium can be obtained the same way it was obtained for equimolar counter diffusion. Thus, from equation (1.97)

$$N_{Az}(z_2 - z_1) = DC_T \ln \frac{C_{B_2}}{C_{B_1}} \tag{1.97a}$$

$$N_{Az}(z - z_1) = DC_T \ln \frac{C_B}{C_{B_1}} \tag{1.97b}$$

Dividing (1.97b) by (1.97a), we get that

$$\frac{z - z_1}{z_2 - z_1} = \frac{\ln \frac{C_B}{C_{B_1}}}{\ln \frac{C_{B_2}}{C_{B_1}}} \tag{1.104}$$

Equation (1.104) can be re-arranged to get that

CHAPTER ONE: FUNDAMENTAL DEFINITIONS AND CONCEPTS

$$C_B = C_{B_1} \left(\frac{C_{B_2}}{C_{B_1}} \right)^{\frac{Z-Z_1}{Z_2-Z_1}} \quad (1.105)$$

Equation (1.105) can, further, be expressed in terms of the diffusing component as follows, knowing that $C_B = C_T - C_A$;

$$C_A = C_T - (C_T - C_{A_1}) \left(\frac{C_T - C_{A_2}}{C_T - C_{A_1}} \right)^{\frac{Z-Z_1}{Z_2-Z_1}} \quad (1.106)$$

Unlike the case of equimolar counter diffusion, the concentration profile, outlined by equation (1.106), is not a straight line. This profile is shown, for one component, in Figure 1.4 below.

Figure 1.4: Concentration Profile for Diffusion through a Stagnant Medium

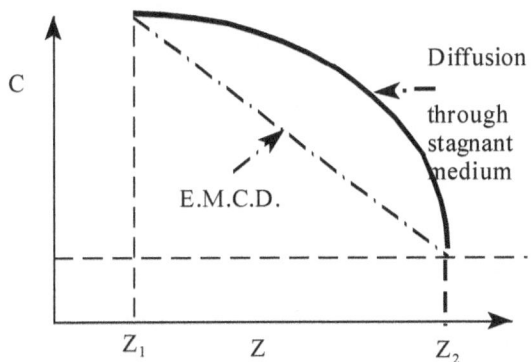

1.4.1.3: The General Case – Neither Equi-molar Counter Diffusion nor Diffusion through a Stagnant Medium

In the region near the gas/liquid interface, the concentration profile for the two components in the gas phase could be illustrated as shown in Figure 1.5 below.

From equations (1.83) and (1.88)

$$N_{A_z} = \frac{N_{AZ} + N_{BZ}}{C_T} C_A + n_{AZ} = (N_{AZ} + N_{BZ}) \frac{C_A}{C_T} - D_{AB} \frac{\partial C_A}{\partial z} \quad (1.107)$$

At steady state, with N_{AZ}, N_{BZ} and D_{AB} constant, equation (1.107) can be rearranged as

$$\int_{C_{A_1}}^{C_{A_2}} \frac{-dC_A}{N_{AZ} C_T - C_A (N_{AZ} + N_{BZ})} = \frac{1}{C_T D_{AB}} \int_{z_1}^{z_2} dz \quad (1.108)$$

and integrated to give, knowing that $z = z_2 - z_1$,

$$N_{A_z} = \frac{N_{A_z}}{N_{AZ} + N_{BZ}} \frac{D_{AB} C_T}{z} \ln \frac{\dfrac{N_A}{(N_{AZ} + N_{BZ})} - \dfrac{C_{A_2}}{C_T}}{\dfrac{N_A}{(N_{AZ} + N_{BZ})} - \dfrac{C_{A_1}}{C_T}} \quad (1.109)$$

Figure 1.5: Interface Concentration Profiles

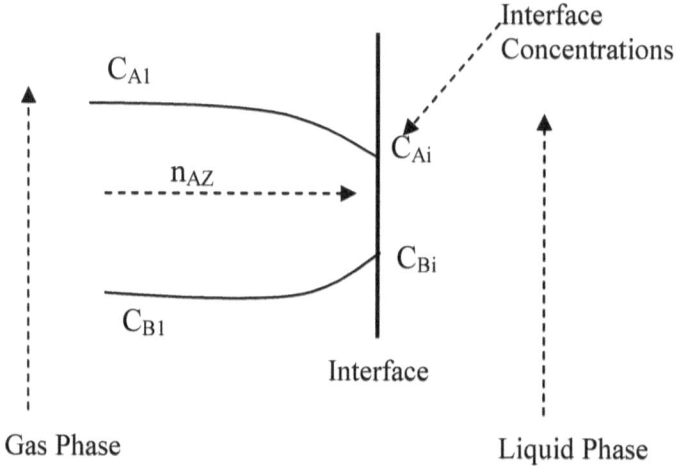

For gases, knowing from equation (1.03) that $C_A = p_A/RT$, the equivalent of equation (1.109) is

$$N_{A_z} = \frac{N_{A_z}}{N_{AZ} + N_{BZ}} \frac{D_{AB} P_T}{z RT} \ln \frac{\dfrac{N_A}{(N_{AZ} + N_{BZ})} - \dfrac{P_{A_2}}{P_T}}{\dfrac{N_A}{(N_{AZ} + N_{BZ})} - \dfrac{P_{A_1}}{P_T}} \quad (1.110)$$

1.4.1.4: Diffusion through Variable Area

Often, the diffusion of material through variable surface area arises in the processing of cylindrically shaped and drop like materials. We shall consider the simplest case of equimolar counter diffusion through a sphere, as shown in the figure below.

For equimolar counter diffusion,
$$N_{A_r} = n_{A_r} \qquad \text{from equation (1.86)}$$
From Fick's law,

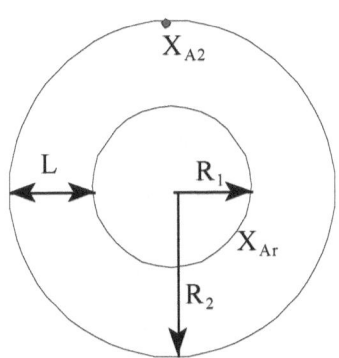

$$n_{A_r} = -DC_T \frac{dX_{A_r}}{dr} \qquad \text{from equation (1.41)}$$

Because of varying area in the radial direction, along which mass diffusion occurs, N_{Ar} is not constant. Instead, $N_{Ar}S_r$ is. S_r is the area, at any radius, r, that is perpendicular to the direction of mass flux. Thus, from (1.86) and (1.41);

$$N_{A_r} S_r = -4\pi r^2 DC_T \frac{dX_{A_r}}{dr}$$
$$= constant = (N_A S)_r \qquad (1.111)$$

Integrating and re-arranging equation (1.111);

$$(N_A S)_r \int_{r_1}^{r_2} \frac{dr}{r^2} = -4\pi DC_T \int_{X_{A_1}}^{X_{A_2}} dX_A \qquad (1.112)$$

33

That is: $(N_A S)_r \left(\dfrac{1}{r_1} - \dfrac{1}{r_2}\right) = 4\pi DC_T (X_{A_1} - X_{A_2})$ (1.113)

from which $(N_A S)_r = DC_T . 4\pi \dfrac{r_1 r_2}{r_2 - r_1}(X_{A_1} - X_{A_2})$ (1.114)

Since the geometric mean radius is defined as

$$r_{gm} = \sqrt{r_1 r_2} \quad (1.115)$$

then $(N_A S)_r = \dfrac{DC_T}{L} . 4\pi r_{gm}^2 (X_{A_1} - X_{A_2})$

$= \dfrac{DC_T}{L} . S_{gm} (X_{A_1} - X_{A_2})$ (1.116)

where S_{gm} is the geometric mean area.

We can, now, evaluate the molar flux from any of our fixed surfaces 1 or 2 as follows. Let us take the flux from surface 1. Then, since from (1.111), $(N_A S)_r = (N_A S)_1$

$$N_{A_1} = \dfrac{DC_T}{L} \dfrac{S_{gm}}{S_1}(X_{A_1} - X_{A_2}) = \dfrac{DC_T}{L} \dfrac{r_2}{r_1}(X_{A_1} - X_{A_2}) \quad (1.117)$$

1.4.2: Diffusion in Turbulent Flow

Analysis of mass transfer in turbulent flow in fluids is, usually, done in two ways, depending on the objectives of the mass transfer operation. In one type of analysis, the boundary layer region and the region of fully developed turbulence in the bulk of the fluid are considered simultaneously. Since the boundary layer region consists of three layers (the laminar sub-layer, the buffer layer and the turbulent layer) instead of the single, steadily growing boundary layer encountered in laminar flow, it is easier to define an effective laminar film thickness such that the resistance to molecular diffusion in this thickness is the same as that offered to mass transfer by these boundary layers mentioned above. In the gas phase, this effective film thickness is designated as z_G while in the liquid phase, it is denoted by z_L.

In the other type of analysis, only the mass transfer in the turbulent bulk, or portions, of the fluid is considered as being dominant. Such situations arise during mass transfer in stirred tanks or vessels. In this case, the eddy diffusivity, ε_D, is introduced together with molecular diffusivity, D and depending on the degree of turbulence, ε_D .may be so much greater than D that it replaces it in the mass transfer equations or may, in fact, tend to infinity.

1.4.2.1: Diffusion in the Gas Phase in the Boundary Layer Regions in Turbulent Flow

Molecular diffusion, in turbulent flow, across the gas interface into the bulk of the gas, may be visualized by looking at the variation of partial pressure of the diffusing gas with distance from the interface into the bulk of the gas. This is illustrated in the diagram below.

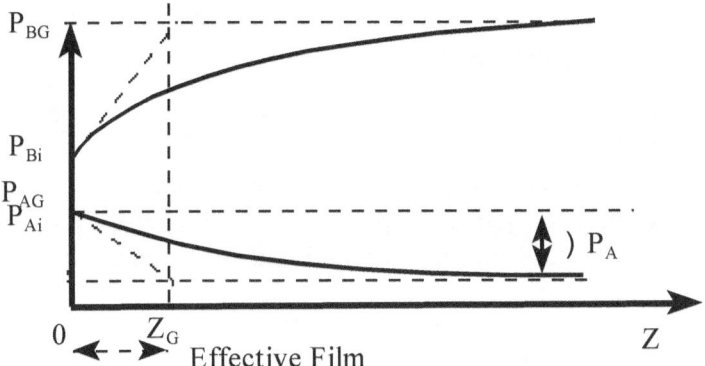

The effective film thickness, Z_G, is defined such that the resistance to molecular diffusion, within it, is equal to the sum of those offered to mass transfer by the real laminar, buffer and turbulent boundary layers combined. Graphically it is determined as the point, on the z-axis, where a straight line from the interface partial pressure (or concentration) cuts the asymptote to the infinite portion of the partial pressure or concentration profile curve of the system. In some cases, this straight line may be a tangent from the interface concentration or partial pressure point of the concentration or partial pressure curve respectively. The asymptote is determined by the value of the bulk partial pressure or concentration of the diffusing component such as is shown for the components, A and B in the diagram above. The reason for this will become apparent when theories of mass transfer at an interface are discussed later on in this book.

For the mean time, we, merely, replace the L for laminar diffusion by z_G such that the equations for equi-molar counter diffusion and diffusion through a stagnant medium become, respectively,

$$N_A = \frac{D_{AB}}{RT z_G}(P_{Ai} - P_{AG}) = k'_G(P_{Ai} - P_{AG}) \qquad (1.118)$$

for equimolar counter diffusion and

$$N_A = \frac{D_{AB}}{RT z_G}(P_{Ai} - P_{AG})\frac{P_T}{P_{BM}} = k_G(P_{Ai} - P_{AG}) \qquad (1.119)$$

for diffusion through a stagnant medium.

Equations (1.118) and (1.119) define the *mass transfer coefficient*, k'_G, for equi-molar counter diffusion and, k_G, for diffusion through a stagnant medium. That is

$$k'_G = \frac{D_{AB}}{RT z_G} \quad \text{and} \quad k_G = \frac{D_{AB}}{RT z_G}\frac{P_T}{P_{BM}} = k'_G \frac{P_T}{P_{BM}} \qquad (1.120)$$

It should be noted that, for ideal gases, the mass transfer coefficient based on partial pressure driving force, k_G, that based on mole fraction driving force, k_y, and that based on concentration driving force, k_C, are related as;

$$k'_G = \frac{k'_y}{P_T} = \frac{k'_C}{RT} \qquad (1.121)$$

1.4.2.2: Diffusion in the Liquid Phase in the Boundary Layer Regions in Turbulent Flow

For liquids, the equivalent equations for equi-molar counter diffusion and diffusion through a stagnant medium are, respectively,

$$N_A = \frac{D_{AB}}{z_L}(C_{AL} - C_{Ai}) = k'_L(C_{AL} - C_{Ai}) \qquad (1.122)$$

for equi-molar counter diffusion and

$$N_A = \frac{D_{AB}}{z_L}(C_{AL} - C_{Ai})\frac{C_T}{C_{BM}} = k_L(C_{AL} - C_{Ai}) \quad (1.123)$$

for diffusion through a stagnant medium. k'$_L$ and k$_L$ are the liquid *mass transfer coefficients*, given by

$$k'_L = \frac{D_{AB}}{z_L} \quad \text{and} \quad k_L = \frac{D_{AB}}{z_L}\frac{C_T}{C_{BM}} = k'_L \frac{C_T}{C_{BM}} \quad (1.124)$$

for equimolar counter diffusion and diffusion through a stagnant medium, respectively. It is, also, to be noted that for liquids, the mass transfer coefficient based on concentration driving force, k_L, that based on liquid mole fraction driving force, k_x, are related as;

$$k'_L = \frac{k'_x}{C_T} \quad (1.125)$$

for equi-molar counter-diffusion, and

$$k_L = \frac{k_L}{C_T} \quad (1.126)$$

for diffusion through a stagnant medium.

These mass transfer coefficients can be related to each other as follows:

$$k'_L = k_L \frac{C_{BM}}{C_T} = k_X x_{BM} \quad (1.127)$$

It is usual to estimate these mass transfer coefficients from dimensionless groups and correlations such as the ones shown below as equations (1.128) and (1.129).

$$\frac{k'd}{D} = \frac{f}{2} \operatorname{Re} Sc^{\frac{1}{3}} \quad (1.128)$$

$$j_D = \frac{k'_G P_T}{G} \psi(Sc) = \psi'(\operatorname{Re}) \quad (1.129)$$

1.4.2.3: Diffusion in the Turbulent Region

While molecular diffusion is due to statistical motion on a molecular scale (Brownian motion), turbulence is statistical motion on a macro scale and can be due to the action of an external, deliberate action.

In turbulent transport

Heat flux = *Eddy thermal diffusivity x Temperature Gradient* (1.130a)

Mass flux = *Eddy mass diffusivity x Concentration Gradient* (1.130b)

Momentum flux = *Eddy momentum diffusivity x Velocity Gradient* (1.130c)

These diffusivities are, usually, represented by the following symbols: α_H = eddy thermal diffusivity
ε_D = eddy mass diffusivity
ε_v = eddy momentum diffusivity

REFERENCES

1. Class Notes in Mass Transfer, Imperial College of Science & Technology, London, 1967-1969
2. http://en.wikipedia.org/wiki/Mass_transfer", April, 2008
3. Analogies Between Heat, Mass, And Momentum Transfer, Wikipedia, April 2008

CHAPTER TWO
THEORIES OF MASS TRANSFER AT AN INTERFACE

2.1: Formulating Mass Transfer Theories

For a theory to be useful in the study and application of mass transfer principles and practice, certain conditions need to be met. These are:
1. There exists a model of the physical conditions at the interface
2. This model can be expressed in mathematical form
3. The resulting mathematical formulation should lead to some solution which enables the evaluation of the mass flux, the mass transfer coefficient or, at least, the effects of the major variables on the transfer process
4. the combination of the physical model and its mathematical expression must be general enough to deal with the major concerns of the mass transfer process

2.2: Basic Definitions and Concepts

2.2.1: Mass Flux

In the mass transfer between two points, say from point 1 to point 2, in a fluid with a bulk velocity, U_Z in the z direction, the total molar flux of component A would be given, from the definitions in Chapter 1, as

$$N_{A_z} = U_Z C_A + n_{A_z} + n'_{A_z} \qquad (2.1)$$

where C_A = molar concentration of A, kmol/m^3
$U_Z C_A$ = convective flux, kmol/m^2.s
n_{AZ} = molar flux of A by molecular diffusion in the z direction, kmol/m^2.s
n'_{AZ} = molar flux of A by eddy diffusion in the z direction, kmol/m^2.s

Since

$$n_{A_z} = -D_{AB}\frac{\partial C_A}{\partial z} = -D_{AB}C_T \frac{\partial X_A}{\partial z} \qquad (2.2)$$

and

$$n'_{A_z} = -\varepsilon_D \frac{\partial C_A}{\partial z} = -\varepsilon_D C_T \frac{\partial X_A}{\partial z} \quad (2.3)$$

$$N_{A_z} = U_z C_A - (D_{AB} + \varepsilon_D)\frac{\partial C_A}{\partial z}$$

$$= U_z C_T X_A - (D_{AB} + \varepsilon_D)C_T \frac{\partial X_A}{\partial z} \quad (2.4)$$

where C_T = total concentration, kmol/m^3
D_{AB} = molecular diffusivity of A through the medium, B, m^2/s
X_A = mole fraction of A
ε_D = eddy diffusivity of A through B, m^2/s

2.2.2: Molecular and Eddy Diffusion

Consider a stationary medium for which U_Z = 0. Then, from equation (2.4)

$$N_{A_z} = -(D_{AB} + \varepsilon_D)\frac{\partial C_A}{\partial z} = -(D_{AB} + \varepsilon_D)C_T \frac{\partial X_A}{\partial z} \quad (2.5)$$

Equation (2.5) illustrates, because of the linear addition of their diffusivities, that molecular diffusion and eddy diffusion occur in parallel. By considering the concentration-distance curve, such as that shown below, it can, also, be seen that eddy diffusivity is a function of the distance from the interface at which mass transfer occurs while molecular diffusivity (the result of Brownian motion) is constant.

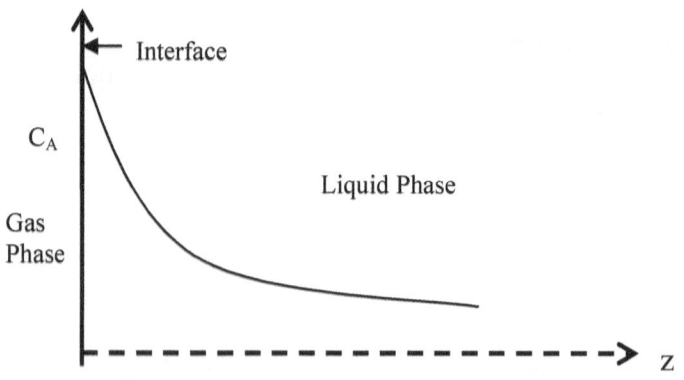

These enable us to classify mass transfer theories by the value of ε_D at the interface.

1. If $\varepsilon_D = 0$ in the vicinity of the interface

$$N_{A_z} = n_{A_z} = -D_{AB}\frac{\partial C_A}{\partial z} = -D_{AB}C_T\frac{\partial X_A}{\partial z} \qquad Fick's\ Law\ (2.2)$$

That is, molecular diffusion controls the mass transfer.

2. If $\varepsilon_D > D_{AB}$, even at the vicinity of the interface, eddy diffusivity controls the mass transfer
3. If $\varepsilon_D \approx D_{AB}$, molecular and eddy diffusion are, simultaneously, important in the mass transfer.

2.2.3: When Molecular Diffusion Controls the Mass Transfer

2.2.3.1: Binary Mixtures

In this case,
$$N_{A_z} = U_z C_T X_A + n_{A_z} = (N_{A_z} + N_{B_z})X_A + n_{A_z} \qquad (2.6)$$

Under conditions of zero net flux, such as in equimolar counter diffusion, $N_{AZ} + N_{BZ} = 0$ and

$$N_{A_z} = n_{A_z} = -D_{AB}\frac{\partial C_A}{\partial z} = -D_{AB}C_T\frac{\partial X_A}{\partial z} \qquad (2.7)$$

2.2.3.2: Multicomponent Mixtures

The molar flux in the z-direction, for each component, i, is given by
$$N_{i_z} = U_z C_T X_i + n_{i_z} \qquad (2.8)$$

the concentration gradient by

$$C_T \frac{dX_i}{dz} = \sum_{j=1}^{n} \frac{1}{D_{ij}}(X_i N_{jz} - X_j N_{iz}) \qquad (2.9)$$

and the molecular diffusivity by

$$D_{i\,mean} = \frac{1}{\sum_{j \neq i}^{n-1} \frac{y_j}{D_{ij}}} \qquad (2.10)$$

2.2.3.3: Consequences of the Conservation of Mass in Molecular Diffusion Control

The mass balance over an element in which mass transfer occurs is given by (see Appendix I)

$$\frac{\partial C_A}{\partial \theta} = -div.N_A + R_A \qquad (2.11)$$

where R_A is the rate of formation of A per unit volume by chemical reaction and θ is time. Substituting equation (2.6) with equation (2.2) into equation (2.11) and simplifying, we get equation (2.12) below such that, in only the z-direction,

$$\frac{\partial C_A}{\partial \theta} + U_z \frac{\partial C_A}{\partial z} = D_{AB} \frac{\partial^2 C_A}{\partial z^2} - C_A \frac{\partial U_z}{\partial z} + R_A \qquad (2.12)$$

where, in three dimensions,

$$div.P = \left[\frac{\partial P_x}{\partial x} + \frac{\partial P_y}{\partial y} + \frac{\partial P_z}{\partial z} \right] \quad P \text{ can be any function} \qquad (2.13)$$

For a dilute solution in which no chemical reaction occurs

$$\frac{\partial C_A}{\partial \theta} = D \left[\frac{\partial^2 C_A}{\partial z^2} \right] \qquad (2.14)$$

This is the mathematical statement of Fick's second law of molecular diffusion. It is, also, known as the diffusion equation and applies, even in bulk flow, when the solution is so dilute that C_A and change of C_A with z are negligible. It is, consequently, a basic equation to all molecular diffusion theories of mass transfer.

2.3: Solutions of the Diffusion Equation and the Associated Mass Transfer Theories

Most theories of mass transfer at an interface are based on a mathematical solution of the diffusion equation and the deduction of the properties, characteristics and consequent utility of the solution. These solutions depend on the kind of assumptions made in solving the diffusion equation. On the face of it, three kinds of assumptions, which may be made in order to solve the diffusion equation, quickly come to mind.

CHAPTER TWO: THEORIES OF MASS TRANSFER AT AN INTERFACE

1. there is steady state molecular diffusion of the diffusing component through a stagnant medium
2. there is unsteady state molecular diffusion into a stationary, semi-infinite medium
3. there is unsteady state molecular diffusion into a stationary, bounded medium.

It is obvious that these are not the only assumptions which can be made. However, professional practice tends to limit the possibilities to systems and processes of commercial or well known value or utility.

2.3.1: The Assumption of Steady State Molecular Diffusion through a Stagnant Medium – The Whitman Film Theory

The main postulations of the theory are outlined below. The assumptions of the physical model are

a. the bulk of the fluid is in turbulent motion
b. there is a film of thickness, L, on the interface, which represents the effective thickness in which mass transfer occurs by molecular diffusion. This film may be stagnant or in plug flow.
c. This film is so thin that the concentration profile is established instantaneously.
d. The bulk of the fluid, outside this film, is turbulent and offers no resistance to mass transfer (i.e. $\varepsilon_D = \infty$ for $z > L$)
e. All resistance to mass transfer is concentrated in the film and is diffusional in nature

<u>Solution of the Diffusion Equation</u>
The mathematical solution of the diffusion equation, on the basis of these assumptions, is presented below.

For steady state molecular diffusion through a stagnant medium, equation (2.14) reduces to

$$D\left[\frac{\partial^2 C_A}{\partial z^2}\right] = 0 \quad \text{since} \frac{\partial C_A}{\partial \theta} = 0 \qquad (2.15)$$

The boundary conditions are

1. At $z = z_1$ (interface or any diffusion starting point), $C_A = C_{A1}$

2. At $z = z_2$ (end of effective film thickness), $C_A = C_{A2}$

Integrating equation (2.15) twice, substituting the boundary conditions above and simplifying, we get that the concentration profile is

$$C_A = C_{A_1} - \frac{C_{A_1} - C_{A_2}}{z_2 - z_1}(z - z_1) \qquad (2.16)$$

Molar Flux

The molar flux is obtained from equations (2.7) and (2.16) as

$$N_{A_z} = n_{A_z} = -D\frac{dC_A}{dz} = D\left[\frac{C_{A_1} - C_{A_2}}{z_2 - z_1}\right] \qquad (2.17)$$

Equation (2.17) represents the mathematical formulation of the Whitman film theory. Note that $z_2 - z_1 = L$, the effective film thickness.

Mass Transfer Coefficient

The mass transfer coefficient, K, is given by

$$K = \left[\frac{N_A}{C_{A_1} - C_{A_2}}\right] = \frac{D}{L} \qquad (2.18)$$

Note that K is proportional to D to the first power and inversely proportional to L, also to the first power. In practice, K is found to be proportional to D raised to the 2/3 power. Thus, the Whitman film theory over-estimates the mass transfer coefficient.

A useful assumption, at this stage, in the Whitman film theory, is that the effective film thickness established on the basis of concentration (mass transfer), L_D, is equal to that established in turbulent flow as a result of velocity gradient (momentum transfer), L_v which is a function of the Reynold's number.

The concentration profile, as explained by the Whitman film theory, is illustrated below

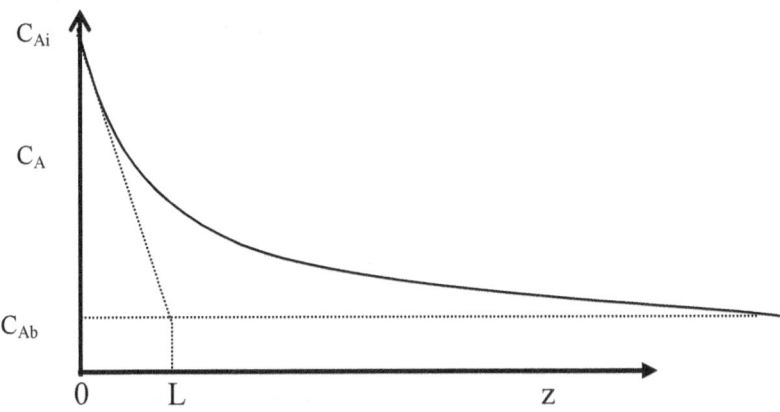

C_{Ai} is the concentration of A at the interface, C_{Ab}, the concentration of A in the bulk of the fluid.

Surface Resistance

The Whitman film theory handles surface resistance, R_S, in the interface as follows. Note that $R_S = 1/K_S$ where K_S is the surface mass transfer coefficient while the clean interface resistance, $R_L = 1/K_L$ where K_L is the mass transfer coefficient based on the Whitman effective film for mass transfer (D/L).

Schematically, the concentration profile in a system having interfacial surface resistance is shown below. When there is no surface resistance, $R_S = 0$, and

$$N_A = K_L(C_{A_i} - C_{A_b}) \quad \text{where } K_L = \frac{D}{L} \qquad (2.19)$$

When there is surface resistance, $R_S \neq 0$, and since N_A is the same through both resistances,

$$N_A = K_S(C_{A_i} - C_{A_S}) = K_L(C_{A_S} - C_{A_b}) \qquad (2.20)$$

Equation (2.20) is, easily, rearranged to obtain

$$N_A = \frac{1}{\frac{1}{K_S} + \frac{1}{K_L}}(C_{A_i} - C_{A_b}) = \frac{1}{R_S + R_L}(C_{A_i} - C_{A_b}) \qquad (2.21)$$

For a soluble surface active agent (surfactant), R_S is of the order of 2000 s/m while for an insoluble surfactant, it is of the order 10^5

s/m. For the absorption of CO_2 in a stirred tank, K_L is of the order of 2×10^{-5} m/s or an $R_L \approx 5 \times 10^4$ s/m.

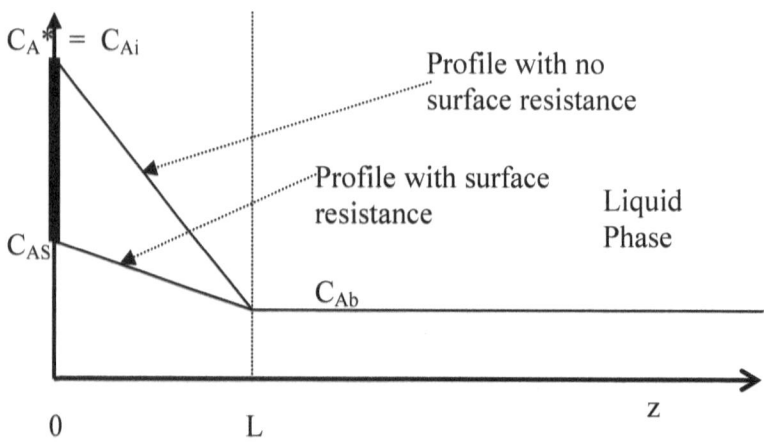

Modifications to the Whitman Film Theory Model

The assumption that $L_D = L_v$ is valid only if the Schmidt number, Sc, is unity. That is

$$Sc = 1 = \frac{v}{D} \qquad (2.22)$$

We can rewrite the expression for the mass transfer coefficient. This enables us to relate the film thickness for mass transfer to the film thickness in momentum transfer which is a function of the Reynold's number. Thus

$$K_L = \frac{D}{L_D} = \frac{D}{L_v} \cdot \frac{L_v}{L_D} \qquad (2.23)$$

From the third power law

$\varepsilon_D \, \alpha \, z^3$ for mass eddy diffusivity

$\varepsilon_v \, \alpha \, z^3$ for momentum eddy diffusivity $\qquad (2.24)$

We may assume that

$\varepsilon_D = \varepsilon_v = a z^3$ where a is a constant $\qquad (2.25)$

provided that this assumption leads to a result that can be confirmed by experimental measurements. A plot of this power law is shown below.

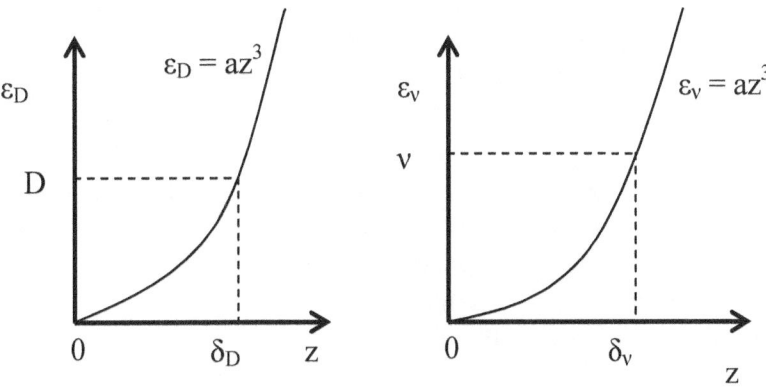

It illustrates the well known boundary layer theory observation that the mass eddy diffusivity, ε_D, becomes equal to the molecular diffusivity, D, at a certain boundary layer thickness denoted by δ_D. Similarly, the momentum eddy diffusivity, ε_v, becomes equal to the momentum diffusivity, v, at the boundary layer thickness, δ_v. That is

$$\varepsilon_v = v \text{ at } z = \delta_v$$
$$\varepsilon_D = D \text{ at } z = \delta_D \qquad (2.26)$$

If we make the reasonable assumption that

$$\frac{\delta_v}{\delta_D} = \frac{L_v}{L_D} \qquad (2.27)$$

Then

$$\frac{\delta_v}{\delta_D} = \left(\frac{v}{D}\right)^{1/3} = Sc^{1/3} = \frac{L_v}{L_D} \qquad (2.28)$$

From equations (2.22) and (2.28), we get that

$$K_L = \frac{D}{L_v}\cdot\left(\frac{v}{D}\right)^{1/3} = \frac{D^{2/3} v^{1/3}}{L_v} \qquad (2.29)$$

K_L is seen to vary with D to the 2/3 power which is in line with experimental observations thus confirming the validity of the assumptions stated in equations (2.25) and (2.27).

2.3.2: The Assumption of Unsteady State Molecular Diffusion into a Stationary Semi-infinite Medium (Zero Net Flux) – The Penetration Theory

The main assumption of the penetration theory is that of a semi-infinite medium. The physical model is that hydrodynamic conditions exist which allow mass transfer by non-steady molecular diffusion. It is assumed that a constant concentration is, instantaneously, established at the interface and that the significant depth of penetration is smaller than the depth of undisturbed liquid. Solute concentration is low so that zero net flux conditions are approximated. Practical possibilities for this model may be found in mass transfer in drops, in stirred tanks and in wetted wall columns.

The mathematical formulation follows from the solution of the diffusion equation (2.14)

$$\frac{\partial C_A}{\partial \theta} = D\left[\frac{\partial^2 C_A}{\partial z^2}\right] \qquad (2.14)$$

with the following boundary conditions:

1	$\theta = 0$	$z \geq 0$	C_A	=	C_{A0}
2	$\theta \geq 0$	$z = 0$	C_A	=	C_{Ai}
3	$\theta \geq 0$	$z = \infty$	C_A	=	C_{A0}

(2.30)

These conditions require that the time of contact be less than a certain value (instantaneous build up of concentration) for the solution to be valid. They show, also, that this is a case of a step concentration function applied to the interface.

Solution of the Diffusion Equation

The solution is very much simplified by the use of a dimensionless concentration defined as follows

$$\Delta = \frac{C_A - C_{A_0}}{C_{A_I} - C_{A_0}} \qquad (2.31)$$

The concentration of A becomes transformed into

$$C_A = C_{A_0} + (C_{A_I} - C_{A_0})\Delta \qquad (2.32)$$

So that

$$\frac{\partial C_A}{\partial \theta} = (C_{A_I} - C_{A_0})\frac{\partial \Delta}{\partial \theta} \qquad (2.33)$$

and
$$\frac{\partial^2 C_A}{\partial z^2} = (C_{A_i} - C_{A_0})\frac{\partial^2 \Delta}{\partial z^2} \qquad (2.34)$$

The diffusion equation is then
$$\frac{\partial \Delta}{\partial \theta} = D_{AB}\frac{\partial^2 \Delta}{\partial z^2} \qquad (2.35)$$

The initial and boundary conditions transform to

1	$\theta = 0$	$z \geq 0$	$\Delta = 0$
2	$\theta \geq 0$	$z = 0$	$\Delta = 1$
3	$\theta \geq 0$	$z = \infty$	$\Delta = 0$

(2.36)

Taking the Laplace transform of both sides of equation (2.35) where Ψ is the Laplace transform of Δ, equation (2.35) becomes

$$\frac{\partial \Psi}{\partial \theta} = D_{AB}\frac{\partial^2 \Psi}{\partial z^2} \qquad (2.37)$$

where
$$\frac{\partial \Psi}{\partial \theta} = \int_0^\infty e^{-p\theta}\frac{\partial \Delta}{\partial \theta}d\theta = p\Psi - \Delta(0) = p\Psi \qquad (2.38)$$

$$\frac{\partial^2 \Psi}{\partial z^2} = \int_0^\infty e^{-p\theta}\frac{\partial^2 \Delta}{\partial z^2}d\theta = \frac{\partial^2}{\partial z^2}\int_0^\infty e^{-p\theta}\Delta d\theta = \frac{\partial^2 \Psi}{\partial z^2} = \frac{d^2 \Psi}{dz^2} \qquad (2.39)$$

Assuming that Ψ and its derivatives are continuous functions, then, from equations (2.37), (2.38) and (2.39), we get

$$\frac{d^2 \Psi}{dz^2} - \frac{p\Psi}{D_{AB}} = 0 \qquad (2.40)$$

which has the solution

$$\Psi = Ae^{z\sqrt{\frac{p}{D_{AB}}}} + Be^{-z\sqrt{\frac{p}{D_{AB}}}} \qquad (2.41)$$

The Laplace transform of the boundary conditions 2 and 3 are

| 2 | $z = 0$ | $\Delta = 1$ | $\Psi = 1/p$ | (2.42) |
| 3 | $z = \infty$ | $\Delta = 0$ | $\Psi = 0$ | |

Applying boundary condition 2 to equation (2.41)

$$\frac{1}{p} = B \qquad (2.43)$$

Applying boundary condition 3 to equation (2.41)
$$0 = A.\infty + B.0 \tag{2.44}$$
from which we deduce that A must be equal to zero so that equation (2.41) becomes

$$\Psi = \frac{1}{p} e^{-z\sqrt{\frac{p}{D_{AB}}}} \tag{2.45}$$

From Laplace Transform Tables, equation (2.45) becomes

$$\Delta = \frac{C_A - C_{A_0}}{C_{A_i} - C_{A_0}} = erfc \frac{z}{2\sqrt{\theta D_{AB}}} \tag{2.46}$$

From which we get that

$$C_A = C_{A_0} + (C_{A_i} - C_{A_0}) erfc \frac{z}{2\sqrt{\theta D_{AB}}} \tag{2.47}$$

This represents the concentration profile shown below.

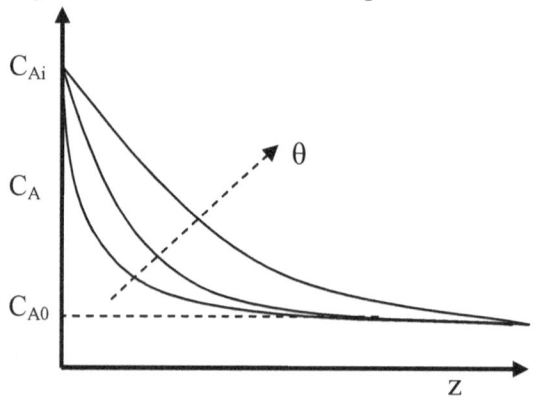

Instantaneous Molar Flux

From equation (2.7)
$$N_A(\theta,z) = n_A(\theta,z) = -D_{AB} \frac{\partial C_A}{\partial z} \tag{2.7}$$

$$\frac{\partial C_A}{\partial z} = (C_{A_i} - C_{A_0}) \frac{\partial}{\partial z}\left(erfc \frac{z}{2\sqrt{\theta D_{AB}}} \right)$$

$$= (C_{A_i} - C_{A_0}) \times \frac{2}{\sqrt{\pi}} e^{-\frac{z^2}{4\theta D_{AB}}} \times \frac{1}{2\sqrt{\theta D_{AB}}} \tag{2.48}$$

CHAPTER TWO: THEORIES OF MASS TRANSFER AT AN INTERFACE

so that

$$N_A(\theta, z) = \sqrt{\frac{D_{AB}}{\pi \theta}} \cdot (C_{A_i} - C_{A_0}) \exp\left(-\frac{z^2}{4\theta D_{AB}}\right) \quad (2.49)$$

The mass flux at the interface, where $z = 0$, is given by

$$N_A(\theta, 0) = \sqrt{\frac{D_{AB}}{\pi \theta}} \cdot (C_{A_i} - C_{A_0}) \quad (2.50)$$

The Mass Transfer Coefficient

Since *Mass Transfer Coefficient* $(K_L) = \dfrac{Flux}{Driving\ Force}$

The instantaneous mass transfer coefficient, $K_L(\theta)$, at the interface, would be given by

$$K_L(\theta) = \frac{N_{A_i}(\theta)}{C_{A_i} - C_{A_0}} = \sqrt{\frac{D_{AB}}{\pi \theta}} \quad (2.51)$$

Note that at $\theta = 0$, $K_L(\theta)$ is infinite. Between $\theta = 0$ and a few micro-seconds later, especially, for pure liquids, the diffusion process is rate controlled. After these few micro-seconds, equation (2.51) takes over. A plot of $N_{Ai}(\theta)$ versus θ is illustrated in the figure below.

If θ_C is the time of contact, the average molar flux over this time is given by

$$N_{A_i} = \frac{1}{\theta_C} \int_0^{\theta_C} N_{A_i}(\theta) d\theta = \frac{1}{\theta_C} \sqrt{\frac{D}{\pi}} \cdot (C_{A_i} - C_{A_0}) \int_0^{\theta_C} \frac{d\theta}{\sqrt{\theta}} \quad (2.52)$$

Since

$$\frac{d\sqrt{\theta}}{d\theta} = \frac{1}{2\sqrt{\theta}} \quad (2.53)$$

$$N_{A_i} = \frac{2}{\theta_C} \sqrt{\frac{D}{\pi}} \cdot (C_{A_i} - C_{A_0}) \int_0^{\theta_C} d(\sqrt{\theta}) = 2\sqrt{\frac{D}{\pi \theta_C}} \cdot (C_{A_i} - C_{A_0}) \quad (2.54)$$

$$= 2 \times Final\ Ins\tan\tan eous\ Value \quad (2.55)$$

Thus, according to the penetration theory,

$$K_L = 2 \cdot \sqrt{\frac{D}{\pi \theta_C}} \quad i.e.\ K \propto D^{0.5} \quad (2.56)$$

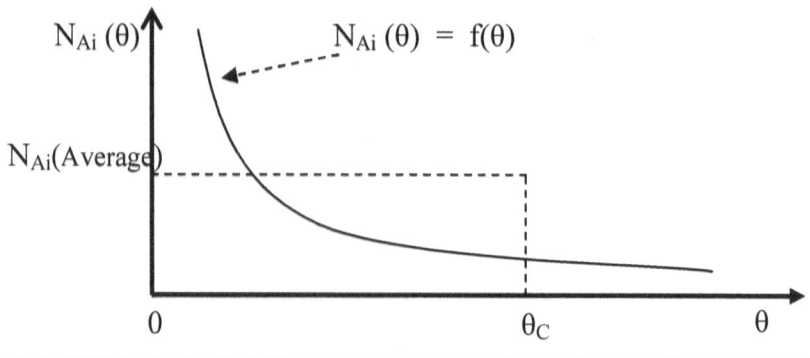

Interfacial Resistance

When interfacial resistance is present, the boundary conditions to the diffusion equation (2.14)

$$\frac{\partial C_A}{\partial \theta} = D\left[\frac{\partial^2 C_A}{\partial z^2}\right] \qquad (2.14)$$

become

1. $\theta = 0$ $\quad 0 < z < \infty \quad C_A = C_{A0}$
2. $\theta = 0$ $\quad z = \infty \quad\quad C_A = C_{A0}$ $\qquad (2.57)$
3. $\theta \geq 0$ $\quad z = 0 \quad\quad K_S(C_{A_i} - C_A) = -D_{AB}\dfrac{\partial C_A}{\partial z}$

The solution was very much simplified by the use of a dimensionless concentration defined as follows

$$\Delta = \frac{C_A - C_{A_0}}{C_{A_I} - C_{A_0}} \qquad (2.31)$$

The concentration of A was transformed into

$$C_A = C_{A_0} + (C_{A_I} - C_{A_0})\Delta \qquad (2.32)$$

So that

$$\frac{\partial C_A}{\partial \theta} = (C_{A_I} - C_{A_0})\frac{\partial \Delta}{\partial \theta} \qquad (2.33)$$

and

$$\frac{\partial^2 C_A}{\partial z^2} = (C_{A_I} - C_{A_0})\frac{\partial^2 \Delta}{\partial z^2} \qquad (2.34)$$

The diffusion equation was then

CHAPTER TWO: THEORIES OF MASS TRANSFER AT AN INTERFACE

$$\frac{\partial \Delta}{\partial \theta} = D\frac{\partial^2 \Delta}{\partial z^2} \qquad (2.35)$$

The initial and boundary conditions can, now, be transformed to

$$
\begin{array}{llll}
1 & \theta = 0 & z \geq 0 & \Delta = 0 \\
2 & \theta \geq 0 & z = \infty & \Delta = 0 \\
3 & \theta \geq 0 & z = 0 & \dfrac{\partial \Delta}{\partial z} = -\dfrac{K_S(1-\Delta)}{D}
\end{array} \qquad (2.58)
$$

Taking the Laplace transform of both sides of equation (2.35) where Ψ is the Laplace transform of Δ, equation (2.35) became

$$\frac{\partial \Psi}{\partial \theta} = D\frac{\partial^2 \Psi}{\partial z^2} \qquad (2.37)$$

where

$$\frac{\partial \Psi}{\partial \theta} = \int_0^\infty e^{-p\theta}\frac{\partial \Delta}{\partial \theta}d\theta = p\Psi - \Delta(0) = p\Psi \qquad (2.38)$$

$$\frac{\partial^2 \Psi}{\partial z^2} = \int_0^\infty e^{-p\theta}\frac{\partial^2 \Delta}{\partial z^2}d\theta = \frac{\partial^2}{\partial z^2}\int_0^\infty e^{-p\theta}\Delta d\theta = \frac{\partial^2 \Psi}{\partial z^2} = \frac{d^2 \Psi}{dz^2} \qquad (2.39)$$

Assuming that Ψ and its derivatives are continuous functions, then, from equations (2.37), (2.38) and (2.39), we got

$$\frac{d^2 \Psi}{dz^2} - \frac{p\Psi}{D} = 0 \qquad (2.40)$$

which has the solution

$$\Psi = Ae^{z\sqrt{\frac{p}{D_{AB}}}} + Be^{-z\sqrt{\frac{p}{D_{AB}}}} \qquad (2.41)$$

The Laplace transform of the boundary conditions 2 and 3 are

$$
\begin{array}{lll}
2 & z = \infty & \Psi = 0 \\
3 & z = 0 & \dfrac{d\Psi}{dz} = -\dfrac{K_S}{D}\left(\dfrac{1}{p} - \Psi\right)
\end{array} \qquad (2.59)
$$

Applying boundary condition 2 to equation (2.41)
$$0 = A.\infty + B.0 \qquad (2.44)$$

from which we deduce that A must be equal to zero so that equation (2.41) becomes

$$\Psi = Be^{-z\sqrt{\frac{p}{D_{AB}}}} \qquad (2.60)$$

Applying boundary condition 3 to equation (2.41), remembering that $z = 0$

$$\frac{d\Psi}{dz} = -B\sqrt{\frac{p}{D}} \cdot e^{-z\sqrt{\frac{p}{D}}} = -B\sqrt{\frac{p}{D}} = -\frac{K_S}{D}(1-B) \qquad (2.61)$$

This gives the value of B as

$$B = \frac{K_S/D}{K_S/D + \sqrt{p/D}} \qquad (2.62)$$

Substituting equation (2.62) in equation (2.60)

$$\Psi = \frac{K_S/D}{K_S/D + \sqrt{p/D}} \cdot e^{-z\sqrt{p/D}} \qquad (2.63)$$

From Laplace Transform Tables, equation (2.63) becomes

$$\Delta = \frac{C_A - C_{A_0}}{C_{A_i} - C_{A_0}}$$

$$= erfc\frac{z}{2\sqrt{\theta D}} - \exp\left(\frac{K_S}{D}z + \frac{K_S^2}{D}\theta\right) \cdot erfc\left(\frac{z}{2\sqrt{\theta D}} + \frac{K_S}{\sqrt{D/\theta}}\right) \qquad (2.64)$$

From which we get the concentration profile as follows in equation (2.65)

$$C_A = C_{A_0} + (C_{A_i} - C_{A_0})\, erfc\frac{z}{2\sqrt{\theta D}}$$

$$- \exp\left(\frac{K_S}{D}z + \frac{K_S^2}{D}\theta\right) \cdot erfc\left(\frac{z}{2\sqrt{\theta D}} + \frac{K_S}{\sqrt{D/\theta}}\right) \qquad (2.65)$$

<u>Instantaneous Molar Flux When Interfacial Resistance is Present</u>

From equations (2.7) and (2.31), at $z = 0$

CHAPTER TWO: THEORIES OF MASS TRANSFER AT AN INTERFACE

$$N_{Ai}(\theta) = -D\frac{\partial C_A}{\partial z} = -D(C_{A_i} - C_{A_0})\frac{\partial \Delta}{\partial z}$$

$$= -D(C_{A_i} - C_{A_0})\left[\begin{array}{c}-\dfrac{2}{\sqrt{\pi}}\cdot\dfrac{1}{2\sqrt{\theta D}} - \dfrac{K_S}{D}e^{\frac{K_S^2\theta}{D}}erfc\left(\dfrac{K_S}{\sqrt{D/\theta}}\right) \\ + e^{\frac{K_S^2\theta}{D}}\cdot\dfrac{2}{\sqrt{\pi}}\cdot e^{-\frac{K_S^2\theta}{D}}\cdot\dfrac{1}{2\sqrt{\theta D}}\end{array}\right]$$

$$= K_S\cdot e^{\frac{K_S^2\theta}{D}}erfc\left(\dfrac{K_S}{\sqrt{D/\theta}}\right)\cdot(C_{A_i} - C_{A_0}) \qquad (2.66)$$

The Instantaneous Mass Transfer Coefficient When Interfacial Resistance is Present

Since *Mass Transfer Coefficient (K)* $= \dfrac{Flux}{Driving\ Force}$

The instantaneous mass transfer coefficient, $K(\theta)$, at the interface, would be given by

$$K(\theta) = \dfrac{N_{A_i}(\theta)}{C_{A_i} - C_{A_0}} = K_S\cdot e^{\frac{K_S^2\theta}{D}}erfc\left(K_S\sqrt{\dfrac{\theta}{D}}\right) \qquad (2.67)$$

In the limit, by L'hospital's rule,

$$\text{Limit}_{\theta\to\infty} K(\theta) = K_S\ \text{Limit}_{\theta\to\infty}\dfrac{erfc\left(K_S\sqrt{\dfrac{\theta}{D}}\right)}{e^{\frac{K_S^2\theta}{D}}}$$

$$= K_S\ \text{Limit}_{\theta\to\infty}\left\{\dfrac{-\dfrac{2}{\sqrt{\pi}}\cdot e^{-\frac{K_S^2\theta}{D}}\cdot\dfrac{K_S}{\sqrt{D}}\cdot\dfrac{1}{2\sqrt{\theta}}}{-\dfrac{K_S^2}{D}\cdot e^{\frac{K_S^2\theta}{D}}}\right\}$$

$$= \sqrt{\dfrac{D}{\theta}}\quad if\ K_S\sqrt{\dfrac{\theta}{D}} > 2\quad or\quad \theta > \dfrac{4D}{K_S^2} \qquad (2.68)$$

That is, the mass transfer coefficient tends to the value given by the penetration theory as θ tends to infinity.

57

2.3.3: The Assumption of Unsteady State Molecular Diffusion into a Stationary Bounded Medium (Zero Net Flux) – The Film Penetration Theory

This theory assumes the hydrodynamic conditions that
a. there is non-steady molecular diffusion into a stagnant medium of finite thickness with the concentration at both boundaries, being kept constant. The assumption of short times of contact, as in the penetration theory, is not made.
b. The solute concentration is low so that zero net flux conditions are well approximated.

From equation (2.14)

$$\frac{\partial C_A}{\partial \theta} = D\left[\frac{\partial^2 C_A}{\partial z^2}\right] \qquad (2.14)$$

The boundary conditions are

1. $\theta = 0$ $0 < z < L$ $C_A = C_{A0}$
2. $\theta > 0$ $z = 0$ $C_A = C_{Ai}$ (Constant)
3. $\theta > 0$ $z = L$ $C_A = C_{A0}$ (2.69)

The solution of equation (2.14) with initial and boundary conditions (2.69) may be obtained by means of the Laplace transform or by the separation of variables. The Laplace transform solution is

$$N_{A_i}(\theta) = \sqrt{\frac{D}{\pi \theta}}.(C_{A_i} - C_{A_0})\left[1 + 2\sum_{n=0}^{\infty} \exp\left(-\frac{n^2 L^2}{D\theta}\right)\right] \qquad (2.70)$$

This converges rapidly for short times. The separation of variables solution is

$$N_{A_i}(\theta) = \frac{D}{L}.(C_{A_i} - C_{A_0})\left[1 + 2\sum_{n=1}^{\infty} \exp\left(-\frac{n^2 \pi^2 D\theta}{L^2}\right)\right] \qquad (2.71)$$

Equations (2.71) and (2.72) are identical if

$$\frac{D\theta}{L^2} = \frac{1}{\pi} \qquad (2.72)$$

$D\theta/L^2$ represents a kind of Fourier number for mass transfer. Schematically

CHAPTER TWO: THEORIES OF MASS TRANSFER AT AN INTERFACE

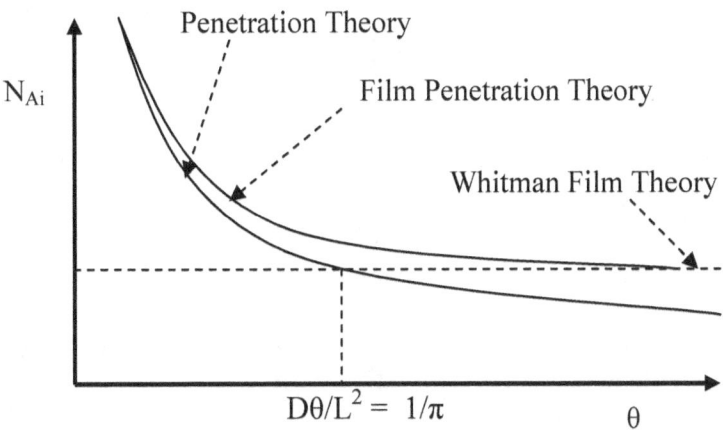

This shows that the penetration theory may be used only if $D\theta/L^2 < 1/\pi$. Then

$$N_{A_i}(\theta) = \sqrt{\frac{D}{\pi \theta}} \cdot (C_{A_i} - C_{A_0}) \left[1 + 2e^{-\frac{L^2}{D\theta}} \right] \quad (2.73)$$

To use the Whitman film theory, $D\theta/L^2 > 1/\pi$, and

$$N_{A_i}(\theta) = \frac{D}{L} \cdot (C_{A_i} - C_{A_0}) \left[1 + 2e^{-\frac{\pi^2 D\theta}{L^2}} \right] \quad (2.74)$$

When $\theta = L^2/\pi D$, an undesirable state of affairs, any of the film penetration theory equations (Laplace transform or separation of variables solutions) may be used. The mass transfer coefficient is a function of time and varies to the ½ power with D for short times and to the first power with D for long times.

2.4: Theories of Mass Transfer Incorporating Eddy Diffusion

2.4.1: Kishenevsky – Kafarov Theory

<u>The Physical Model</u>

The hydrodynamic conditions which exist allow for non-steady mass transfer by molecular and eddy diffusion across the interface. Interface concentration is constant and established instantaneously. The effective depth of penetration of solute is smaller than the thickness of the surface elements. The surface is renewed by large eddies, of uniform and constant concentration, which arrive from the bulk of the fluid. Residence times at the surface vary over s sufficiently narrow range that they can be represented by an average contact time, τ. The eddy diffusivity within the eddies is constant and independent of the distance of the eddies from the surface. Solute concentration is low enough for zero net flux conditions to apply.

Since ε_D is constant, equations (2.14) and the boundary conditions (2.30) still apply except that $C_{A1} - C_{Ai}$ and $C_{A2} = C_{A0}$ now represent the mathematical formulation and D_{AB} is replaced by an effective diffusivity, $D_{eff} = D_{AB} + \varepsilon_D$. That is

$$\frac{\partial C_A}{\partial \theta} = D\left[\frac{\partial^2 C_A}{\partial z^2}\right] \qquad (2.14)$$

with
1 $\theta = 0$ $z \geq 0$ $C_A = C_{A0}$
2 $\theta \geq 0$ $z = 0$ $C_A = C_{Ai}$ (2.30)
3 $\theta \geq 0$ $z = \infty$ $C_A = C_{A0}$

The solution, in the usual manner, gives the concentration profile as

$$C_A(\theta, z) = C_{A_0} + (C_{A_i} - C_{A_0}) \, erfc \frac{z}{2\sqrt{\theta D_{eff}}} \qquad (2.75)$$

and the instantaneous molar flux as

Instantaneous Molar Flux

$$N_{A_i}(\theta, 0) = \sqrt{\frac{D_{eff}}{\pi \theta}} \cdot (C_{A_i} - C_{A_0}) \qquad (2.76)$$

Since contact time is assumed to be, approximately, the same for all surface elements, the surface renewal can be represented by the Higbie model (see Chapter Three). Then

$$N_{A_i} = 2 \cdot \sqrt{\frac{D_{eff}}{\pi \tau}} \cdot (C_{A_i} - C_{A_0}) \qquad (2.77)$$

so that the mass transfer coefficient is

The Mass Transfer Coefficient

$$K = 2 \sqrt{\frac{D_{AB} + \varepsilon_D}{\pi \tau}} \qquad (2.78)$$

When there is intensive stirring so that $\varepsilon_D >>> D_{AB}$

$$K = 2 \sqrt{\frac{\varepsilon_D}{\pi \tau}} \propto D^0 \qquad (2.79)$$

2.4.2: Turbulent Transfer Theory

The Physical Model

The hydrodynamic conditions which exist allow for non-steady mass transfer by molecular and eddy diffusion across the interface. Interface concentration is constant and established instantaneously. The effective depth of penetration of solute is smaller than the thickness of the surface elements. Solute concentration is low enough for zero net flux conditions to apply. The bulk concentration is uniform and constant. Over the zone of primary resistance to mass transfer, eddy diffusivity is represented by a power function, $\varepsilon_D = az^n$ with the eddies being damped out at the interface by surface tension.

Mathematical Formulation

From equation (2.4), with $U_z = 0$,

$$N_{A_z} = -(D_{AB} + az^n)\frac{\partial C_A}{\partial z} \qquad (2.80)$$

From equation (2.13), no bulk flow, no chemical reaction

$$\frac{\partial C_A}{\partial \theta} = -\frac{\partial N_A}{\partial z} = \frac{\partial}{\partial z}\left[(D_{AB} + az^n)\frac{\partial C_A}{\partial z}\right] \qquad (2.81)$$

The boundary conditions remain the same as in (2.30)

1 $\theta = 0$ $z \geq 0$ $C_A = C_{A0}$
2 $\theta \geq 0$ $z = 0$ $C_A = C_{Ai}$ (2.30)
3 $\theta \geq 0$ $z = \infty$ $C_A = C_{A0}$

Specific Solutions

If

1. a is small so that $D_{AB} \gg az^n$, $D_{eff} = D_{AB}$, gives the penetration theory
2. $n = 0$, $D_{eff} = D_{AB} + a$ = constant, gives the Kishenevsky-Kafarov theory
3. $n = 1$, $D_{eff} = D_{AB} + az$, solution is similar to that obtained for heat conduction in solids (see Carslaw and Jaeger; Heat Conduction in Solids)
4. $n = \infty$; $D_{eff} = D_{AB} + az^{\infty}$
Solution is obtained by going back to the original model of eddy diffusivity and defining dimensionless groups as follows:

$$Z = \frac{z}{\delta_D}, \quad \overline{\theta} = \frac{D\theta}{\delta_D^2}, \quad D_{AB} = a\delta_D^n \quad (2.82)$$

Substituting equation (2.82) in equation (2.81) we get

$$\frac{\partial C_A}{\partial \overline{\theta}} = \frac{\partial}{\partial Z}\left[(1 + aZ^n)\frac{\partial C_A}{\partial Z}\right] \quad (2.83)$$

For $Z < 1$ as $n \to \infty$,
$$Z^n \to 0 \text{ and } D_{eff} \to D_{AB} \quad (2.84)$$
Gives the conditions for molecular diffusion control
For $Z > 1$ as $n \to \infty$,
$$Z^n \to \infty \text{ and } D_{eff} \to aZ^n \to \infty \quad (2.85)$$
Gives the conditions of the film penetration theory

The Mass Transfer Coefficient

The mass transfer coefficients for the special solutions outlined above vary with time as follows

1. For small a (Penetration theory)

$$K(\theta) = \sqrt{\frac{D_{AB}}{\pi\theta}}. \quad \text{and} \quad Sh(\theta) = \frac{1}{\sqrt{\pi}}\theta^{-\frac{1}{2}} \quad (2.86)$$

2. For $n = 0$ (Kishenevsky – Kafarov theory)

CHAPTER TWO: THEORIES OF MASS TRANSFER AT AN INTERFACE

$$K(\theta) = \sqrt{\frac{D_{AB} + a}{\pi \theta}} \quad \text{and} \quad Sh(\theta) = \sqrt{\frac{1 + \frac{a}{D}}{\pi}} \theta^{-\frac{1}{2}} \quad (2.87)$$

4. For $n = \infty$ (Film Penetration theory)

For $\Theta > \frac{1}{\pi}$, $\quad K(\theta) = \frac{D_{AB}}{\delta_D} \quad$ and $\quad Sh(\theta) = 1 \quad (2.88)$

For $\Theta < \frac{1}{\pi}$, $\quad K(\theta) = \frac{D_{AB}}{\delta_D}, \quad Sh(\theta) = \frac{1}{\sqrt{\pi}} \theta^{-\frac{1}{2}} \quad (2.89)$

where the Sherwood number is defined as

$$Sh(\theta) = K(\theta) \frac{\delta_D}{D_{AB}} \quad (2.90)$$

2.5: WORKED EXAMPLE

Example 1

A liquid flows down, in laminar flow, along the inside of a hollow column up which gas containing some contaminant is blown to reduce the concentration of the contaminant in the gas stream. If the inside diameter of the column is D, derive an expression for the molar flux of the component, designated as A, from the gas into the liquid film formed along the column.

Answer

Consider the column and the detail of the film formed along the insides of a wetted wall column as shown below.

Assume that the sub-layer, formed, is at flooding velocity as the liquid surface moves with velocity, U_S, in plug flow. The transfer of component A in the system described above is then

$$N_{A_i}(\theta) = \sqrt{\frac{D}{\pi \theta}} \cdot (C_{A_i} - C_{A_0}) \quad (2.50)$$

where $\theta = h/U_S$. (1)
The molar flux can, now, be expressed as a function of height of column as

63

$$N_{A_i}(h) = \sqrt{\frac{DU_S}{\pi h}} \cdot (C_{A_i} - C_{A_0}) \qquad (2)$$

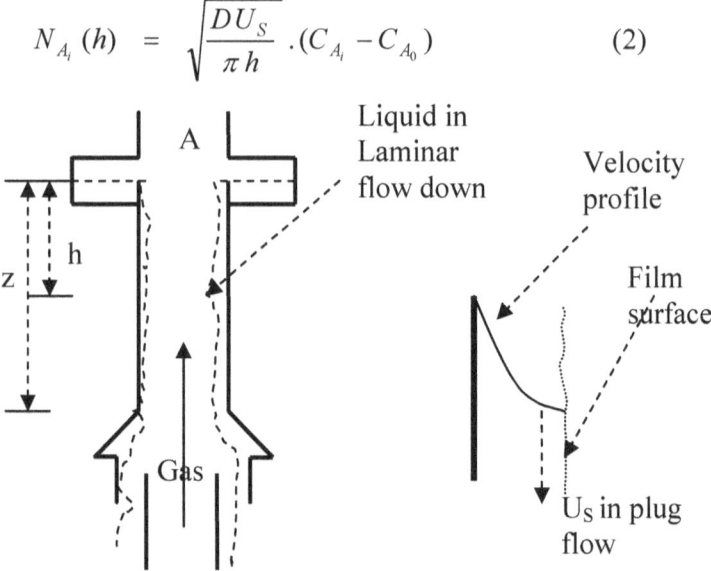

If the rate of absorption of A is designated as W, then
$$dW = N_{A_i}(h) \cdot b \cdot dh \qquad (3)$$
where b is the perimeter of the column. Substituting equation (2) into equation (3) and integrating

$$W = b \int_0^\varepsilon N_{A_i}(h) dh = b \cdot \sqrt{\frac{DU_S}{\pi}} \cdot (C_{A_i} - C_{A_0}) \int_0^\varepsilon \frac{dh}{\sqrt{h}}$$

$$= 2b \cdot \sqrt{\frac{DU_S}{\pi}} \cdot (C_{A_i} - C_{A_0}) \qquad (4)$$

W is the average rate of absorption into the falling film. Thus

$$N_{A_i} = \frac{W}{bz} = 2 \cdot \sqrt{\frac{DU_S}{\pi z}} \cdot (C_{A_i} - C_{A_0}) \qquad (5)$$

Note the similarity between the expression for the molar flux, based on height (equation (2)) to that for the average absorption rate, W (equation (5)).

Usually,

$$U_S = 1.5u; \quad F_V = \frac{V_L}{b} = \text{volumetric peripheral flowrate} \qquad (6)$$

$$V_L = ub\delta \quad \text{where } \delta = \text{film thickness} \qquad (7)$$

CHAPTER TWO: THEORIES OF MASS TRANSFER AT AN INTERFACE

$$F_V = u\,\delta \text{ so that } N_{A_i} = 2 \cdot \sqrt{\frac{1.5\,D\,F_V}{\pi\,z\,\delta}} \cdot (C_{A_i} - C_{A_0}) \tag{8}$$

REFERENCES

1. Kishenevsky M. Kh and Pamfilov A. P.; Zh. Prikl. Khim, 1949, 22, 1173: Kafarov V. V. Fundamentals of Mass Transfer; Vysshaya Shkola, Moscow, 1962, in Russian, referred to in Class Notes, Imperial College of Science & Technology, 1967 – 1969)

2. Whitman W. G., Chem. Met. Eng., 1923, 29, 147
3. Toor H. L. and Marchello J. M.; AIChE (1958), 4, 97
4. King C. J.; Ind. Engrng. Chem. Fundamentals; 1966, 5, 1

CHAPTER THREE
INTERFACIAL PROPERTIES AND PHENOMENA

3.1: Primary and Secondary Properties and Processes at a Mass Transfer Interface

Natural phenomena which occur at an interface during mass transfer may be, arbitrarily, classified as primary if they precede or precipitate other natural or induced phenomena, also, at the interface. Typical of such primary phenomena is surface tension while typical secondary processes are interfacial hydrodynamic instability and interfacial surface renewal. How these affect mass transfer operations is discussed in the following pages.

3.1.1: Why We Need to Understand Interfacial Phenomena and Properties

Consider an interface separating two phases, A and B, all at constant temperature..

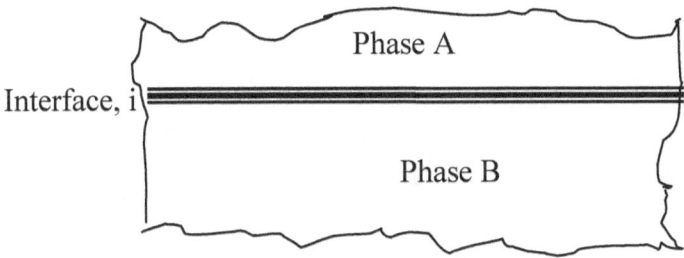

There are many possibilities for A and B as each can either be a liquid, a gas or a vapour or indeed a mixture of these. Each could, also, be stationary or be in motion in laminar or turbulent flow. Suppose, however, for simplicity of analysis, we assume that phase A is a gas and phase B is a liquid and that both phases are stationary. Under these conditions, the Sherwood numbers, Sh, for the gas and liquid phases, A and B, will, normally be given as

$$Sh_A = const. \operatorname{Re}_A^{m_A} \operatorname{Re}_B^{n_A} . Sc_A^{r_A} \qquad (3.1)$$

$$Sh_B = const. \operatorname{Re}_B^{m_B} \operatorname{Re}_A^{n_B} . Sc_B^{r_B} \qquad (3.2)$$

where Re is the Reynold's number (= $\rho du/\mu$), Sc, the Schmidt number (= $\mu/\rho D$) and ρ is the fluid density, μ the fluid viscosity, u the fluid velocity, D its the molecular diffusivity and d the characteristic diameter of the conduit.

The Reynold's number, Re, can be neglected for both phases since none of them has any velocity. It will also be true that the Reynold's number for the bulk of the liquid will also represent that at the interface. Since this is negligible, it may imply, from equations (3.1) and (3.2) that the interface does not interfere with the mass transfer process.

Consider, however, that it is only the interface which has tension, σ_i, whose value may vary along and near the interface. If we look at the balance of forces at the interface, as illustrated in the diagram below, we can see that nature will continuously endeavour to balance these surface forces. This will lead, with mass transfer going on and hence concentration changes, to continuous surface renewal and which, in turn, leads to continuous change in mass transfer coefficients.

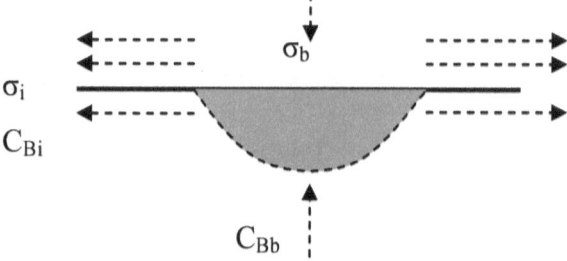

The variation of surface tension with concentration is illustrated in the figure below.

What these go to show is that the Reynold's number, Re, will no longer be representative of conditions at the interface and that interfacial phenomena will affect the magnitude of interfacial area, S. Thus, as W, the rate of mass transfer, is given by

$$W = N_A \cdot S = K(C_{A_i} - C_{A_b})S \qquad (3.3)$$

K and S will be affected by surface tension driven disturbances

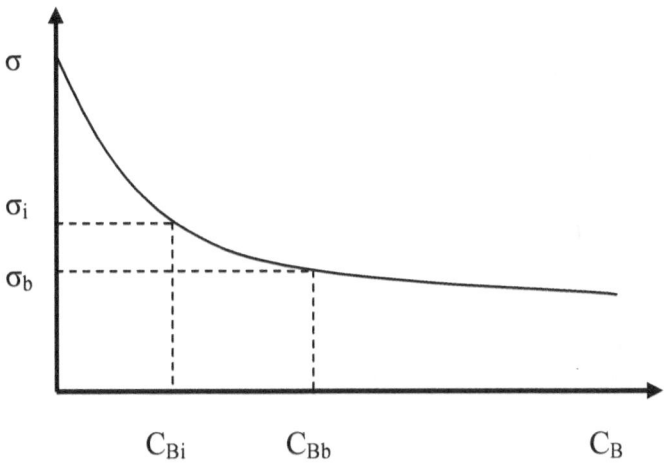

If phase A or phase B or both move, interfacial effects, on mass transfer, are, still, not eliminated since surface instabilities, introduced by the motion, become significant. This will be treated in more detail later.

3.2: Interfacial Properties of Pure Liquids

3.2.1: Surface Tension

Although surface tension is a well known and studied subject, our interest here is in how it affects mass transfer. It is well known that when a pure liquid surface is in contact with a pure gas, that there is a thin surface zone on either side each less than 1.0 μm thick. We know also that pure liquids possess both static surface tension, σ_∞, and dynamic surface tension, σ, both of which are related as

$$\sigma = \sigma_\infty + (\sigma_0 - \sigma_\infty)e^{-\frac{t}{\tau}} \qquad (3.4)$$

where σ_0, is the dynamic surface tension at time t = 0, and τ is the relaxation time, that is the time for σ to approach σ_∞.

Equation (3.4) is known as the relaxation formula. For non-polar molecules, such as CCl_4, the dynamic surface tension is equal to the static surface tension while for polar molecules, such as water, it is not.

To get an idea of the order of magnitude of relaxation times, the relaxation time for the surface tension of pure water is as follows:

$\sigma_0 = 0.180\ N/m;\quad \sigma_\infty = 0.072\ N/m \quad \tau = 10^{-3}\ sec onds,$

3.2.2: Interfacial Tension

Consider a lens of liquid, B, on liquid, A, with which it is immiscible but in contact with air at constant temperature and pressure.

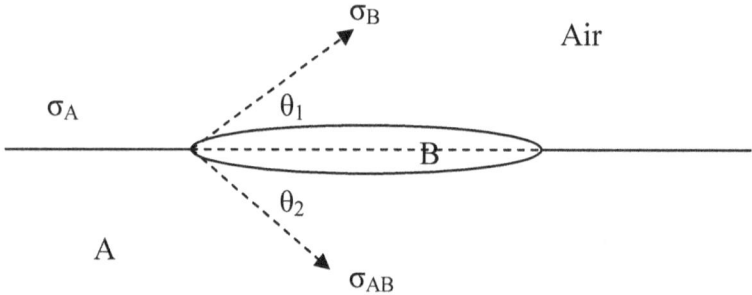

At equilibrium,

$$\sigma_A = \sigma_B \cos\theta_1 + \sigma_{AB}\cos\theta_2 \qquad (3.5)$$

σ_{AB} is the interfacial tension. For spreading of B on A to occur

$$\sigma_A > \sigma_B \cos\theta_1 + \sigma_{AB}\cos\theta_2 \qquad (3.6)$$

As θ_1 and $\theta_2 \to 0$

$$\sigma_A > (\sigma_B + \sigma_{AB}) \qquad (3.7)$$

A spreading coefficient can, then, be defined as

$$S = \sigma_A - (\sigma_B + \sigma_{AB}) \qquad (3.8)$$

so that spreading occurs when S = 0.

CHAPTER THREE: INTERFACIAL PHENOMENA

A very useful relationship between surface and interfacial tension is the so called Antonoff's rule that at S = 0

$$\sigma^*_{AB} = \sigma^*_A - \sigma^*_B \qquad (3.9)$$

where the asterisk indicates that the values are for saturated solutions. Equation (3.9) applies to systems of water and non-polar liquids such as H_2O/CCl_4, si approximately valid for water and slightly polar liquids such as H_2O/C_6H_6, H_2O/C_7H_8, $H_2O/CHCl_3$ but does not apply to water and strongly polar liquids such as water and higher alcohols and for cases in which S < 0 such as H_2O/CS_2.

For a miscible system, σ_{AB} = 0 and
$$S = \sigma_A - \sigma_B \qquad (3.10)$$
Spreading occurs if
$$S > 0 \quad i.e. \quad \sigma_A > \sigma_B \qquad (3.11)$$
Thus, a liquid of lower surface tension will spread on a liquid of higher surface tension. This is known as the Marangoni principle.

3.2.2.1: Interfacial Tension in the Presence of Solutes

The presence of solutes affects the value of the static surface tension. Let us designate static surface tension in the presence of solutes as $\sigma_{S\infty}$ and that of the pure liquid, A, as $\sigma_{A\infty}$.

1. If $\sigma_{S\infty} < \sigma_{A\infty}$, the solute accumulates at the surface and is said to be positively adsorbed at the surface.
2. If $\sigma_{S\infty} > \sigma_{A\infty}$, the solute spreads at the surface and is said to be negatively adsorbed
3. In a binary mixture, one component can be positively adsorbed while the other is negatively adsorbed. A good example is the water/ethanol mixture where the water is negatively adsorbed at the surface while ethanol is positively adsorbed.

The diagrams below illustrate the static surface tension-concentration curves for such systems.

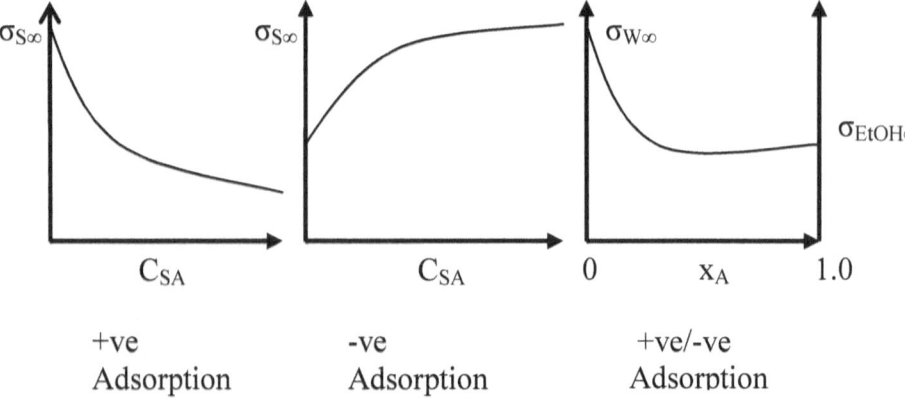

| +ve | -ve | +ve/-ve |
| Adsorption | Adsorption | Adsorption |

3.2.2.2: Interfacial Tension in Distillation

If we designate mole fraction as x, then

1. If $\dfrac{d\sigma_\infty}{dx} < 0$ the more volatile component adsorbs positively at the interface. Such a system is called a surface tension positive system

2. If $\dfrac{d\sigma_\infty}{dx} > 0$ the system is a surface tension negative system

3.2.3: Surface Pressure, Π

Consider a surface illustrated below in which two different regions have different surface pressures, Π_1 and Π_2

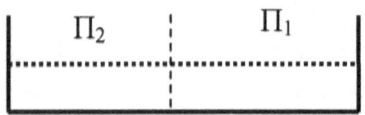

The surface pressure is a suction force but in analogy to pneumatic pressure where $\Delta P = P_1 - P_2$

$$\Delta \Pi = \Pi_1 - \Pi_2 \tag{3.12}$$

If the interface is clean, $\Pi_2 = \Pi_0 = 0$ so that
$$\Delta\Pi = \Pi_1 \qquad (3.13)$$
Surface pressure can be described in terms of suction force as follows

$$\Pi_1 = \frac{F_{S_0} - F_{S_1}}{L} = \sigma_0 - \sigma_{S_1} \qquad (3.14)$$

where F_{S0}, σ_0 and F_{S1}, σ_{S1} are the suction force and surface tension for the pure surface and contaminated surface respectively. Similarly,

$$\Pi_2 = \frac{F_{S_0} - F_{S_2}}{L} = \sigma_0 - \sigma_{S_2} \qquad (3.15)$$

so that
$$\Delta\Pi = \Pi_1 - \Pi_2 = \sigma_{S_2} - \sigma_{S_1} \qquad (3.16)$$

Typically, for a positively adsorbing system, the interfacial tension and the surface pressure vary with surface concentration of contaminant as illustrated below

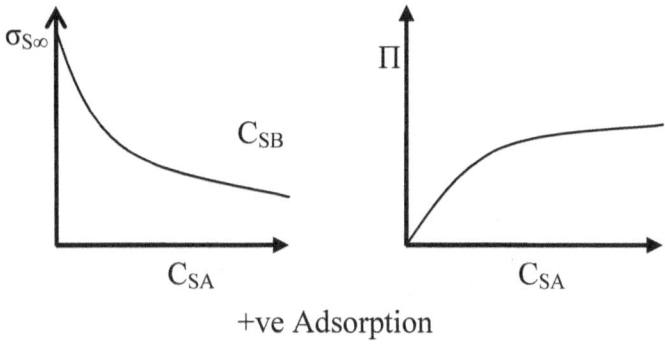

+ve Adsorption

3.2.4: Surface Concentration

If we define

Γ = *molecules per unit area*
Γ_m = *moles per unit area* $\qquad (3.17)$

Consider, in detail, a portion of a surface on which some solute

has been positively adsorbed. The bulk, C, and surface, Γ, concentration layout on the surface is illustrated below

$$\Gamma = \Gamma_b + \Gamma_s \qquad (3.18)$$

where

Γ_S = *surface excess concentration*

$$= -\frac{1}{kT} \cdot \frac{d\sigma_S}{d\ln a} = -\frac{a}{kT} \frac{d\sigma_S}{da} = -\frac{C_{S_A}}{kT} \frac{d\sigma_S}{dC_{S_A}}$$

$$= \frac{C_A}{kT} \frac{d\Pi}{dC_A} \qquad (3.19)$$

where a = activity and $k = R/N_A$, and R is the universal gas constant while N_A is the Avogadro's number for component A.

$$\Gamma_b = C_{S_A} \cdot N_A \cdot d = \text{bulk contribution} \qquad (3.20)$$

$$d = (V_m N_A)^{\frac{1}{3}} \approx 10\,\mu m = \text{molecule diameter} \qquad (3.21)$$

Frequently, Γ_b is negligible.

The area, A, available to one molecule at the interface, not the cross-sectional area, is

$$A = \frac{1}{\Gamma} \qquad (3.22)$$

3.2.5: $\Pi - A$ Diagrams

The characteristics of the several models, which have been proposed to reflect the relationships between surface pressure, Π, and surface area, A, are illustrated below.

3.2.5.1: The Gaseous Film Model

The surface is considered as a gaseous film and the equation, describing it, is modeled after the three dimensional ideal gas equation, PV = nRT. It is, also, assumed that solute concentration in the film is small and that there is no attraction between molecules. Thus, this ideal surface is described by

$$\Pi A = kT \quad (3.23)$$

The ideal surface isotherm is illustrated below.

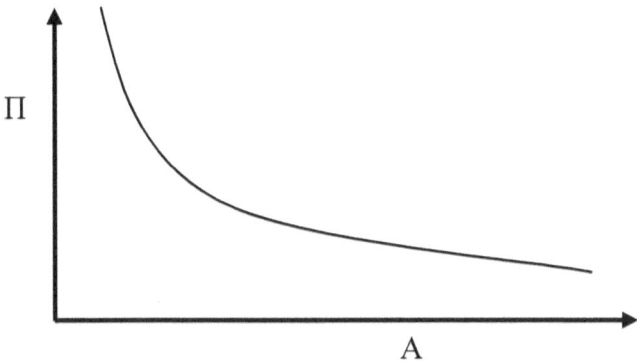

For the case in which

$$\Gamma = \Gamma_S = -\frac{C_{S_A}}{kT}\frac{d\Pi}{dC_{S_A}} \quad (3.19)$$

$$\frac{1}{A} = \frac{C_{S_A}}{kT}\frac{d\Pi}{dA}\frac{dA}{dC_{S_A}} \quad \text{from (3.19) and (3.22)} \quad (3.24)$$

From equation (3.23),

$$\Pi = \frac{kT}{A} \quad \text{from which} \quad \frac{d\Pi}{dA} = -\frac{kT}{A^2} \quad (3.25)$$

Substituting equation (3.25) into equation (3.24)

$$\frac{1}{A} = -\frac{C_{S_A}}{A^2}\frac{dA}{dC_{S_A}} \quad \text{or} \quad \frac{dA}{A} = -\frac{dC_{S_A}}{C_{S_A}}$$

i.e $\ln A = -\ln C_{S_A} + \text{constant}$ or $AC_{S_A} = \text{constant}$ (3.26)

i.e. $\Gamma = \alpha C_{S_A}$ where α is a constant (3.27)

3.2.5.2: The Mobile Film Model

Just like the van der Waal's gas allowed for the molecules of a gas to have some real volume, the mobile film model allows that the molecules in a surface film also occupy some area of the surface. Thus, equation (3.23) becomes modified as shown below

$$\Pi(A - a_m) = kT \quad (3.28)$$

where a_m is the actual area occupied by the molecules. The isotherm is shown below

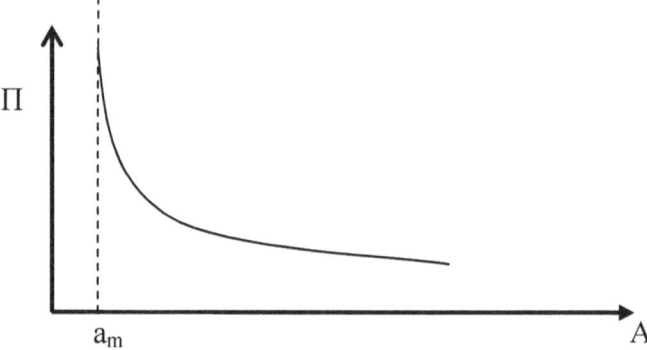

Because this assumption is more realistic than the ideal surface model, it is not surprising that quite a number of systems show this behaviour.

3.2.5.3: The Cohering Film Model

This model, in addition to allowing that the molecules occupy some surface area, also, allows for attraction between molecules.
Since
$$\Pi = \Pi_k + \Pi_S \quad (3.29)$$
where Π_k is the surface pressure defined by the kinetic theory assumptions and

$$\Pi_S = -\frac{B}{A^{\frac{3}{2}}}, \quad \text{where } B \text{ is a constant} \quad (3.30)$$

the surface equation of state is given by

$$(\Pi - \Pi_S)(A - a_m) = kT \quad (3.31)$$

or

$$\left(\Pi + \frac{B}{A^{\frac{3}{2}}}\right)(A - a_m) = kT \qquad (3.32)$$

The isotherm is shown below

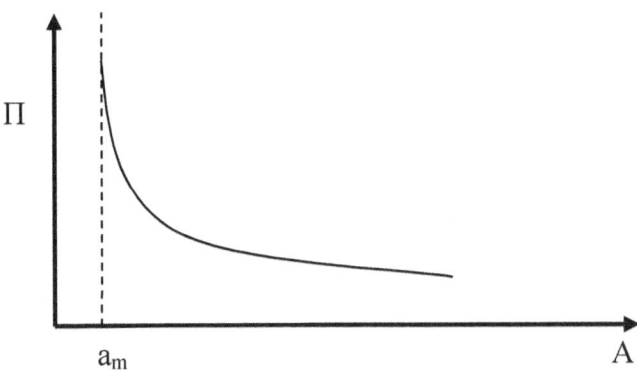

3.2.5.4: The Expanded Liquid Film Model

This model assumes, in addition to finite surface area of molecules and attraction between them, that Π_S is constant and less than zero. The isotherm, shown below, is seen to intersect the A axis at a fixed value

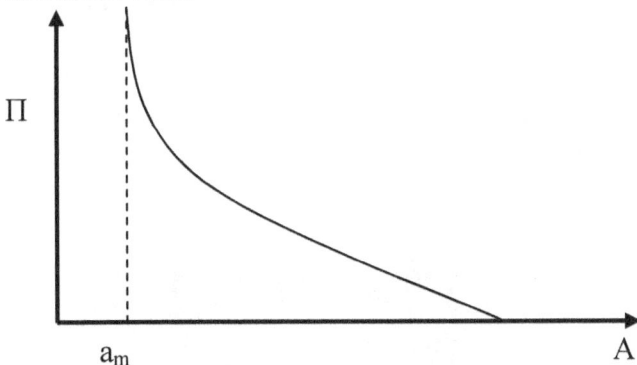

3.2.5.5: The Condensed Film Model

The isotherm, in this model, shows that the surface area is constant and that the condensed film is independent of surface pressure.

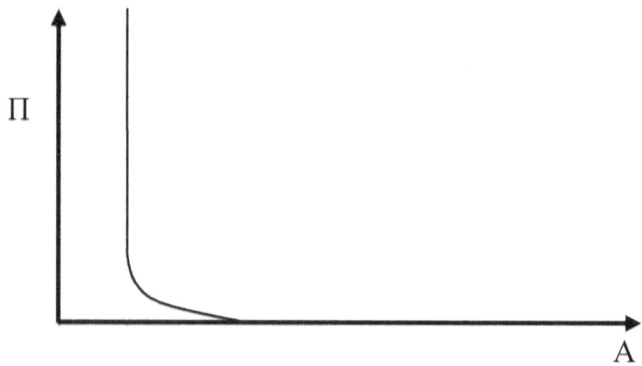

3.2.6: Surface Compressibility

Surface compressibility, γ_S, is the fractional change in surface area, A, per unit change in surface pressure, Π. That is

$$\gamma_S = -\frac{1}{A}\frac{dA}{d(\Delta\Pi)} \qquad (3.33)$$

For a clean interface, $\Delta\Pi = \Pi_1 - \Pi_2 = \Pi_1 = \Pi$ so that

$$\gamma_S = -\frac{1}{A}\frac{dA}{d\Pi} = -\frac{1}{A}\frac{dA}{dC_{S_A}}\frac{dC_{S_A}}{d\Pi} \qquad (3.34)$$

For a gaseous film,

$$\Pi A = kT \qquad (3.23)$$

$$A = \frac{1}{\Gamma} \qquad (3.22)$$

$$\Gamma = \alpha C_{S_A} \quad \text{where } \alpha \text{ is a constant} \qquad (3.27)$$

$$= \frac{1}{A} \qquad \text{from } (3.22)$$

That is: $A = \dfrac{1}{\alpha C_{S_A}}$ and $\dfrac{dA}{dC_{S_A}} = -\dfrac{1}{\alpha C_{S_A}^2}$ (3.35)

From the Gibbs adsorption isotherm

$$\Gamma_S = \frac{C_{S_A}}{kT}\frac{d\Pi}{dC_{S_A}} = \frac{1}{A} \qquad (3.36)$$

So that $\dfrac{dC_{S_A}}{d\Pi} = \dfrac{AC_{S_A}}{kT} = \dfrac{1}{\alpha kT}$ (3.37)

Substituting equations (3.22), (3.35) and (3.37) in equation (3.34)

$$\gamma_S = \alpha C_{S_A} \cdot \frac{1}{\alpha C_{S_A}^2} \cdot \frac{1}{\alpha kT} = \frac{1}{\alpha kT C_{S_A}} \quad (3.38)$$

If the modulus of surface compressibility is denoted by M_{SC}

$$M_{SC} = \frac{1}{\gamma_S} = \alpha kT C_{S_A} \quad (3.39)$$

3.2.7: Surface Viscosity

According to Maxwell's law of relaxation

$$\frac{dR}{dt} = \Phi\eta - \frac{R}{\tau} \quad (3.40)$$

where R = the deformation stress or the deviation of pressure from its equilibrium value (in the general case)
η = the rate of deformation
Φ = a modulus corresponding to the type of deformastion
τ = relaxation time

Equation (3.40) may be re-written as

$$R = \tau\Phi\eta - \tau\frac{dR}{dt} \quad (3.41)$$

which shows that there is time lag before a stressed surface reaches equilibrium stress.

a. <u>For large τ i. e. $\Phi\eta \gg R/\tau$</u>

$$\frac{dR}{dt} = \Phi\eta \quad (3.42)$$

For an elastic material, for example,

$$\eta = \frac{d\varepsilon}{dt} \text{ so that } R = \Phi_\varepsilon \cdot \varepsilon \quad (3.43)$$

where ε is the strain.

b. <u>For small τ i. e. instantaneous attainment of equilibrium</u>

$$\frac{dR}{dt} = 0 \text{ so that } R = \tau\Phi\eta \quad (3.44)$$

For example, in Newtonian flow, R = shear stress and

$$\eta = \frac{dv}{dz} \quad \text{so that} \quad R = \tau\Phi\frac{dv}{dz} = \mu\frac{dv}{dz} \qquad (3.45)$$

Note that the fluid viscosity, μ, is related to the shear relaxation time, τ, and the shear modulus, Φ, as

$$\mu = \tau.\Phi \qquad (3.46)$$

3.3: Interfacial Phenomena

3.3.1: Surface Shear

By analogy with the three dimensional fluid which has nine stress components (six shear stresses and three normal stresses), a surface has four stress components, expressed in terms of surface tension, namely two normal stresses (σ_{yy}, σ_{zz}) and two shear stresses, (σ_{yz}, σ_{zy}).

When deformation occurs by pure shear, the area remains unchanged. In compression or dilation, deformation is assumed to occur similarly though each will have a different τ, Φ, and η. For example, there will be Φ_S, τ_S, μ_S, and η_S, for shear and Φ_C, τ_C, μ_C, and η_C for compression.

Deformation of a surface results in the setting up of stresses which result in dynamic surface tension. Thus

$$\text{Surface Shear}, R, = \sigma_{S_{yz}} \qquad (3.47)$$

Since $\eta = \dfrac{dv}{dz}$ from equation (3.45), the equation for surface flow (non-Newtonian) is then, from equation (3.41), (3.45) and (3.47),

$$\sigma_{S_{yz}} = \tau\Phi\frac{dv}{dz} - \tau\frac{d\sigma_{S_{yz}}}{dt} \qquad (3.48)$$

For Newtonian flow,

$$\sigma_{S_{yz}} = \tau\Phi\frac{dv}{dz} = \mu_S\frac{dv}{dz} \qquad (3.49)$$

μ_S is the surface shear viscosity.

3.3.2: Similarity Deformation

For an isotropic surface, such as a Newtonian surface,
$$R = \sigma_S - \sigma_{S_\infty} \tag{3.50}$$
For a non-isotropic surface,
$$R = \frac{1}{2}(\sigma_{S_{yy}} + \sigma_{S_{zz}}) \tag{3.51}$$
Now
$$\eta = \frac{1}{s}\frac{ds}{dt} = \eta_{yy} + \eta_{zz} = \frac{dv}{dy} + \frac{dw}{dz} \tag{3.52}$$
where s is the area for similarity deformation. Substituting equations (3.50), (3.51) and (3.52) into equation (3.41), remembering that $\sigma_{S\infty}$ is constant,
$$\sigma_S = \sigma_{S\infty} + \tau_C \Phi_C \left(\frac{dv}{dy} + \frac{dw}{dz}\right) - \tau_C \frac{d\sigma_S}{dt} \tag{3.53}$$

For a Newtonian surface, $\mu_C = \tau_C \Phi_C$ and $\dfrac{d\sigma}{dt} = 0$ so that
$$\sigma_S = \sigma_{S\infty} + \mu_C \left(\frac{dv}{dy} + \frac{dw}{dz}\right) \tag{3.54}$$

When these values are to be determined experimentally, the definitions used are
$$\mu_S = \frac{surface\ shear\ stress}{rate\ of\ shear} \tag{3.55}$$
$$\mu_C = \frac{deviation\ of\ surface\ pressure\ from\ static\ value}{rate\ of\ similarity\ deformation} \tag{3.56}$$
Typical experimental values are
$$10^{-5} < \frac{\mu_S}{\mu_C} < 10^{-1}\ sP \tag{3.57}$$
where sP = surface Poise and 1 Poise = 10^{-3} N/m. Note that
$$Bulk\ vis\cos ity,\ \mu = \frac{\mu_S}{\delta_S} \tag{3.58}$$
δ_S can be in the range of 10^{-9} m (1.0 nm). This will put the value of the bulk viscosity in the range $100 < \mu < 10^3$ N/m.

3.3.3: The Effect of Dynamic Surface Tension on Molecular Diffusion at the Interface

It is clear, from the above definitions and analyses, that mass transfer, by molecular diffusion at an interface, will be affected by these surface properties and phenomena. The problem is one of estimating the nature and magnitude of this effect.

The analysis is based on the diffusion equation (2.15)

$$\frac{\partial C_A}{\partial t} = D\left[\frac{\partial^2 C_A}{\partial z^2}\right] \tag{2.15}$$

with boundary conditions

$$\begin{array}{llll} 1 & t = 0 & 0 < z < \infty & C_A = C_{A0} \\ 2 & t \geq 0 & z = \infty & C_A = C_{A0} \\ 3 & t \geq 0 & z = 0 & n_{A_z} + \dfrac{\partial \Gamma_{m_S}}{\partial t} = 0 \end{array} \tag{3.59}$$

Taking the Laplace transform of both sides of equation (2.15) where $\overline{C}_A$ is the Laplace transform of C_A, equation (2.15) becomes

$$\frac{\partial \overline{C}_A}{\partial t} = D\frac{\partial^2 \overline{C}_A}{\partial z^2} \tag{3.60}$$

where

$$\frac{\partial \overline{C}_A}{\partial t} = \int_0^\infty e^{-pt}\frac{\partial C_A}{\partial t}dt = p\overline{C}_A - C_A(0) = p\overline{C}_A - C_{A_0} \tag{3.61}$$

$$\frac{\partial^2 \overline{C}_A}{\partial z^2} = \int_0^\infty e^{-pt}\frac{\partial^2 C}{\partial z^2}dt = \frac{\partial^2}{\partial z^2}\int_0^\infty e^{-pt}C\,dt = \frac{\partial^2 \overline{C}_A}{\partial z^2} = \frac{d^2 \overline{C}_A}{dz^2} \tag{3.62}$$

Assuming that $\overline{C}_A$ and its derivatives are continuous functions, then, from equations (3.60), (3.61) and (3.62), we get

$$\frac{d^2 \overline{C}_A}{dz^2} - \frac{p\overline{C}_A}{D} = -\frac{C_{A_0}}{D} \tag{3.63}$$

with transformed Laplace transform of the boundary conditions

CHAPTER THREE: INTERFACIAL PHENOMENA

$$\begin{array}{lll} 2 & z = \infty & \overline{C}_A = \dfrac{C_{A_0}}{p} \\ & & \\ 3 & z = 0 & \overline{n}_{A_z} + \dfrac{\partial \overline{\Gamma}_m}{\partial t} = 0 \end{array} \qquad (3.64)$$

Where $\overline{n}_{A_z}$ and $\dfrac{\partial \overline{\Gamma}_m}{\partial t}$ are the Laplace transforms of n_{AZ} and $\dfrac{\partial \Gamma_m}{\partial t}$ respectively. Integrating equation (3.63)

$$\overline{C}_A = A e^{z\sqrt{\frac{p}{D}}} + B e^{-z\sqrt{\frac{p}{D}}} + \frac{C_{A_0}}{p} \qquad (3.65)$$

Applying boundary condition 2 to equation (3.65)

$$\frac{C_{A_0}}{p} = A.\infty + B.0 + \frac{C_{A_0}}{p} \qquad (3.66)$$

from which we deduce that A must be equal to zero so that equation (3.65) becomes

$$\overline{C}_A = B e^{-z\sqrt{\frac{p}{D}}} + \frac{C_{A_0}}{p} \qquad (3.67)$$

In order to apply boundary condition 3 to equation (3.65), we must evaluate the Laplace transform of the boundary condition 3, remembering that $z = 0$. Thus

$$\overline{n}_{A_z} = -D\frac{\partial \overline{C}_A}{\partial z} = \text{Laplace transform of } n_{A_z} \qquad (3.68)$$

$$\frac{\partial \overline{\Gamma}_m}{\partial t} = p\overline{\Gamma}_m - \Gamma_m(0) = \text{Laplace transform of } \frac{\partial \Gamma_m}{\partial t} \qquad (3.69)$$

But, $\quad \Gamma = \alpha C_{S_A} \quad \text{where } \alpha \text{ is a constant} \qquad (3.27)$

$$\overline{\Gamma}_m = \alpha \overline{C}_{A_0} \quad \text{where } \alpha \text{ is a constant} \qquad (3.70)$$

where $\overline{\Gamma}$ is the Laplace transform of Γ. From equations (3.69) and (3.70)

$$\frac{\partial \overline{\Gamma}_m}{\partial t} = p\alpha \overline{C}_A \quad \text{since } \Gamma_m(0) = 0 \qquad (3.71)$$

From equations (3.68) and (3.71), the third boundary condition (equation (3.64)) becomes

$$-D\frac{\partial \overline{C}_A}{\partial z} + p\alpha \overline{C}_A = 0 \qquad (3.72)$$

From equation (3.67), with $z = 0$,

$$\frac{d\overline{C}_A}{dz} = -B\sqrt{\frac{p}{D}} \quad (3.73)$$

Also, at $z = 0$, equation (3.67) becomes

$$\overline{C}_A = B + \frac{C_{A_0}}{p} \quad (3.74)$$

Substituting equations (3.74) and (3.73) in equation (3.72)

$$DB\sqrt{\frac{p}{D}} + p\alpha B + p\alpha \frac{C_{A_0}}{p}$$

$$= B\left(p\alpha + D\sqrt{\frac{p}{D}}\right) + \alpha C_{A_0} \quad (3.75)$$

From which we get that

$$B = -\frac{\alpha C_{A_0}}{p\alpha + D\sqrt{\frac{p}{D}}} \quad (3.76)$$

Substituting equation (3.76) into equation (3.67)

$$\overline{C}_A = \frac{C_{A_0}}{p} - \frac{\alpha C_{A_0}}{p\alpha + D\sqrt{\frac{p}{D}}} e^{-z\sqrt{\frac{p}{D}}} \quad (3.77)$$

The inverse of equation (3.77) is

$$C_A = C_{A_0}\left[1 - \exp\left(\frac{z}{\alpha} + \frac{Dt}{\alpha^2}\right).erfc\left(\frac{z}{2\sqrt{Dt}} + \frac{\sqrt{Dt}}{\alpha}\right)\right] \quad (3.78)$$

At $z = 0$

$$C_{Az} = 0 = C_{A_0}\left[1 - \exp\left(\frac{Dt}{\alpha^2}\right).erfc\left(\frac{\sqrt{Dt}}{\alpha}\right)\right] \quad (3.79)$$

Equation (3.79) shows how the subterranean concentration varies with time.

In terms of interfacial tension, remember that, for a gaseous film, surface excess concentration, Γ, is related to bulk concentration, C, as

$$\Gamma = \alpha C_{S_A} \quad \text{where } \alpha \text{ is a constant} \quad (3.27)$$

CHAPTER THREE: INTERFACIAL PHENOMENA

and to interfacial tension as

$$\Gamma_S = -\frac{1}{kT}\cdot\frac{d\sigma_S}{d\ln a} = -\frac{a}{kT}\frac{d\sigma_S}{da} = -\frac{C_{S_A}}{kT}\frac{d\sigma_S}{dC_{S_A}}$$

$$= \frac{C_A}{kT}\frac{d\Pi}{dC_A} \qquad (3.19)$$

At $z = 0$

$$\Gamma_m = \alpha C_{A_{(z=0)}} = 0 = -\frac{C_{A_{(z=0)}}}{RT}\frac{d\sigma_S}{dC_{A_{(z=0)}}}$$

That is $d\sigma_S = -\alpha RT\, dC_{A_{(z=0)}}$ (3.80)

$$dC_{A(z=0)} = -C_{A_0}\left[\frac{D}{\alpha^2}\exp\left(\frac{Dt}{\alpha^2}\right).erfc\left(\frac{\sqrt{Dt}}{\alpha}\right) \right.$$
$$\left. +\frac{-2}{\sqrt{\pi}}\exp\left(\frac{Dt}{\alpha^2}\right).\exp\left(-\frac{Dt}{\alpha^2}\right)\frac{\sqrt{D}}{\alpha}\frac{1}{2\sqrt{t}}\right]dt$$

$$= -C_{A_0}\left[\frac{D}{\alpha^2}\exp\left(\frac{Dt}{\alpha^2}\right).erfc\left(\frac{\sqrt{Dt}}{\alpha}\right) - \frac{1}{\alpha}\sqrt{\frac{D}{\pi t}}\right]dt \quad (3.81)$$

Substituting equation (3.81) into equation (3.80) and integrating

$$\int_{\sigma_\infty}^{\sigma_S}d\sigma_S = \int_0^t \alpha RT C_{A_0}\left[\frac{D}{\alpha^2}\exp\left(\frac{Dt}{\alpha^2}\right).erfc\left(\frac{\sqrt{Dt}}{\alpha}\right) - \frac{1}{\alpha}\sqrt{\frac{D}{\pi t}}\right]dt$$

which gives

$$\sigma_\infty - \sigma_S = \alpha RT C_{A_0}\left(\frac{2}{\alpha}\sqrt{\frac{Dt}{\pi}} - I\right) \qquad (3.82)$$

where $I = \int\left[\frac{D}{\alpha^2}\exp\left(\frac{Dt}{\alpha^2}\right).erfc\left(\frac{\sqrt{Dt}}{\alpha}\right)\right]dt$ (3.83)

I can be obtained by graphical integration and part of it is also given by

$$I = \int_0^x \exp(x^2).erfc\, x\, dx^2 = 2\int_0^x x\exp(x^2).erfc(x)\, dx \quad (3.84)$$

Estimation of the Instantaneous Molar Flux with Dynamic Surface Tension Present

From equation (2.7), at $z = 0$

$$N_A = -D\frac{\partial C_A}{\partial z}\bigg|_{z=0}$$

$$= DC_{A_0}\left[\frac{1}{\alpha^2}\exp\left(\frac{Dt}{\alpha^2}\right).erfc\left(\frac{\sqrt{Dt}}{\alpha}\right) - \frac{1}{\sqrt{\pi Dt}}\right] \quad (3.85)$$

and

$$n_A = \sqrt{\frac{D}{\pi t}}C_{A_0} - \frac{D}{\alpha}C_{A_0}\exp\left(\frac{Dt}{\alpha^2}\right).erfc\left(\frac{\sqrt{Dt}}{\alpha}\right) \quad (3.86)$$

3.3.4: The Marangoni Effect

A liquid of lower surface tension will spread on one of a higher surface tension. The phenomenon was first studied, scientifically, by Lord Thompson. If this spreading and the spreading pressure on an interface are considered,

$$\Pi_i = \sigma_0 - \sigma_i \qquad \Pi_b = \sigma_0 - \sigma_b$$

Interface with film of lower surface tension

$$\longleftarrow \Pi \longrightarrow$$
Surface Pressure

it will be found that, in the Marangoni effect, $\Delta\Pi = \Pi_b - \Pi_i$ is positive. This is a condition for spreading as shown from equation (3.14). $\Pi_i = \sigma_0 - \sigma_i$ at the interface while in the bulk, $\Pi_b = \sigma_0 - \sigma_b$ where σ_0 is the surface tension of the uncontaminated surface. The spreading pressure difference is, therefore $\Delta\Pi = (\Pi_b - \Pi_i) > 0$ \hfill (3.87)

For generality of analysis, three surface tension systems are defined as follows:

Condition	System Type
If $\frac{d\sigma}{dx} > 0$	Surface tension negative system
If $\frac{d\sigma}{dx} = 0$	Surface tension neutral system (3.88)

If $\frac{d\sigma}{dx} < 0$ Surface tension positive system

3.3.4.1: Influence of the Marangoni Effect on the Size of Interfacial Area

1. Supported Interfacial Area

These are interfacial mass transfer areas such as films which are formed on or are supported by a solid surface. Several consequences of the effect of this on their structure, participation or contribution to mass transfer are discussed below.

a. <u>Mass Transfer into Film, e.g. Wetted Wall Column</u>

This is often a surface tension negative system. Interfacial area is unstable for mass transfer into the film. Tendency is towards a decrease in interfacial area. Contraction of the liquid film, as mass transfer progresses, is experienced. The thin film formed has limited capacity for the solute so that its bulk concentration tends to approach the equilibrium value, C_A^*, which becomes eventually the final concentration C_{AF} in the film. The diagrams below illustrate the physical situation, the equilibrium concentration curve and the surface tension-concentration curve, respectively.

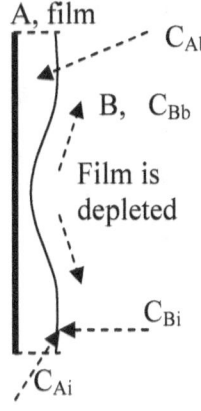
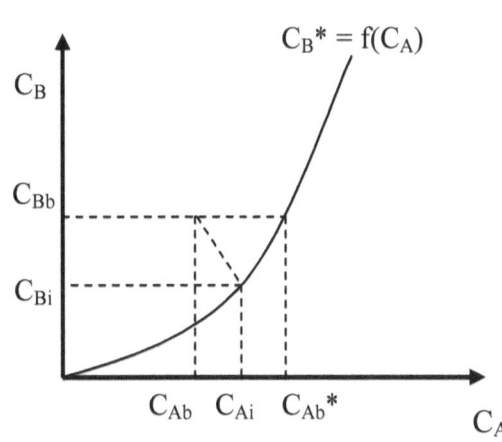

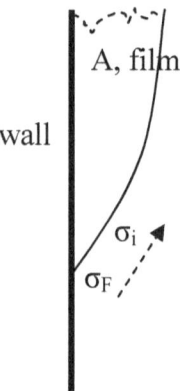

b. <u>Mass Transfer out of Film</u>

This is a surface tension positive system. Interfacial area is stable for mass transfer out of the film and is, therefore, approximately, constant. The physical conditions of the film at the solid wall together with the variation of surface tension with concentration are illustrated below

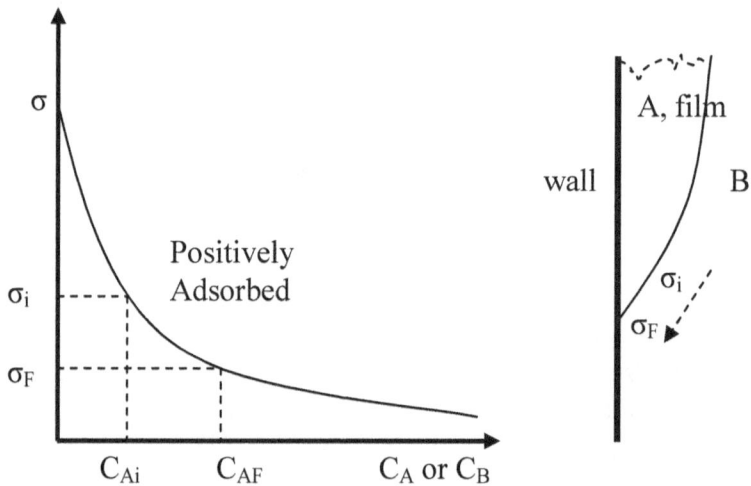

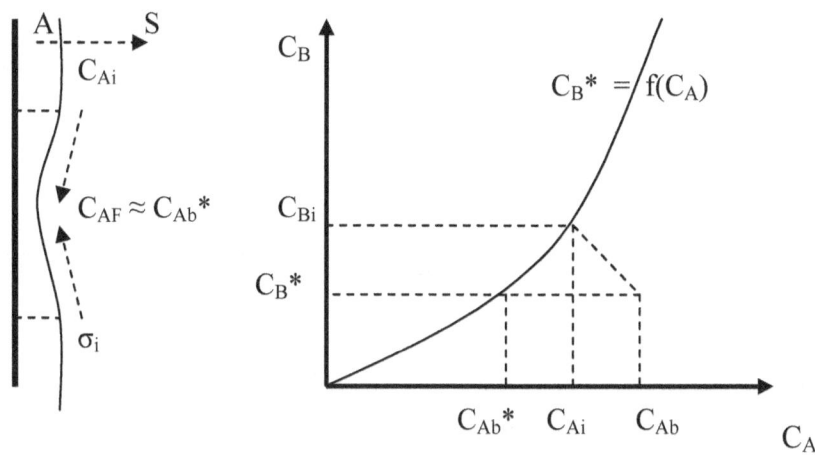

2. Unsupported Interfacial Area

These are interfacial mass transfer areas such as are provided by drops and bubbles which do not rest on or are supported by a solid surface. The consequences of this on their structure, participation or contribution to mass transfer are discussed below.

a <u>Coalescence</u>

Rupture at the edges

When a drop or bubble hits a liquid or gas interface, the action of gravity thins out the liquid or gas film involved and coalescence takes place. There is a rest time during which there is rupture of the drop before coalescence can take place.

When there is no mass transfer, the distribution of rest times shows a normal distribution, with the peak of the frequency distribution curve indicating the most probable rest time.

When there is mass transfer out of the drop and the film capacity for taking in the mass transferred is limited, gravity reinforces the Marangoni effect. Because rest times are very short, coalescence is, practically, instantaneous.

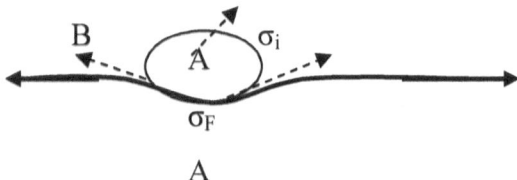

A

When mass transfer is into the drop, the thin film is rapidly depleted. Induced flow opposes coalescence resulting in long rest times. Coalescence is, therefore, more difficult.

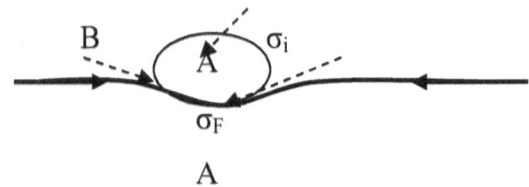

A

For spray towers, where the disperse phase, continually, transforms to the continuous, there is a large amount of coalescence and, thus, decreased interfacial area. If the operation is from the continuous to the disperse phase, there is less coalescence.

In packed columns, supported or unsupported interfacial area can occur depending on which (liquid or bubbles), preferentially, wets the packing. Bubble size distribution is, however, constant no matter the direction of mass transfer since the pores limit the size of the bubbles.

b.   Distillation

The three typical cases are encountered in distillation processes as stated earlier.

CHAPTER THREE: INTERFACIAL PHENOMENA

Condition	System Type	
If $\dfrac{d\sigma}{dx} > 0$	Surface tension negative system	
If $\dfrac{d\sigma}{dx} = 0$	Surface tension neutral system	(3.88)
If $\dfrac{d\sigma}{dx} < 0$	Surface tension positive system	

The variation of surface tension with fractional composition of the heavier phase is illustrated, schematically, below for the three surface tension systems.

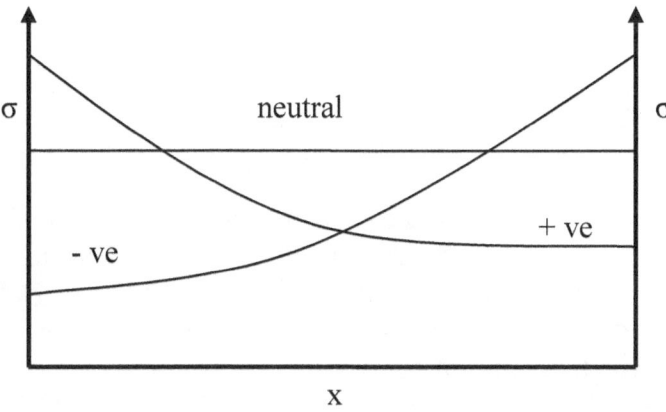

i. Surface Tension Positive System

Coalescence is decreased. There is a tendency to foam formation resulting in large interfacial area and high efficiencies. This occurs at low vapour velocities. For very high vapour velocities, however, even surface tension positive systems will not foam.
The neck of the forming drop becomes strengthened and counteracts drop formation. This results in low plate efficiency.

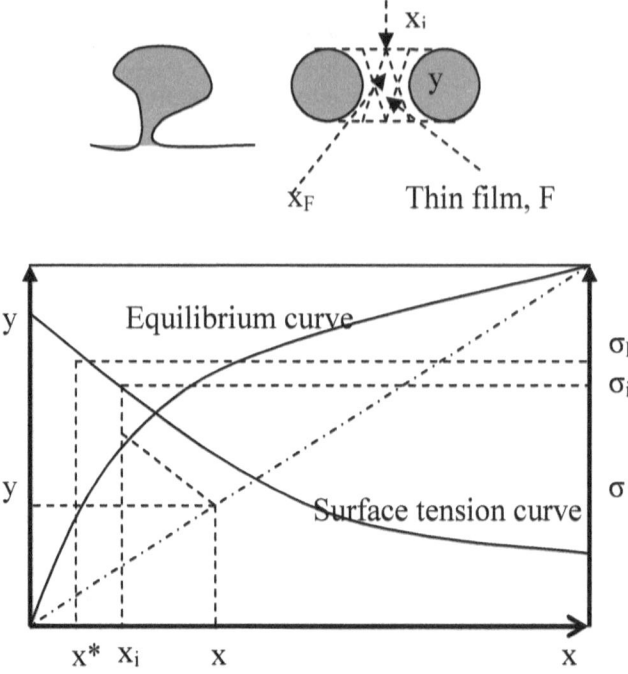

The diagrams above illustrate, first, the physical conceptualisation and, secondly, the graphical variation of surface tension, σ, and vapour composition, y, with liquid composition, x, of the diffusing component in the situation in which drops or bubbles interact with a liquid film.

ii. <u>Surface Tension Neutral Systems</u>

Nothing spectacular happens

iii. <u>Surface Tension Negative Systems</u>

There is no tendency to foam. Spray is formed. The neck of the forming drop gets broken so that drop formation occurs, creating more interfacial area. This results in high plate efficiency. The vapour composition, y, and surface tension, σ, versus liquid composition, x, curves are shown below

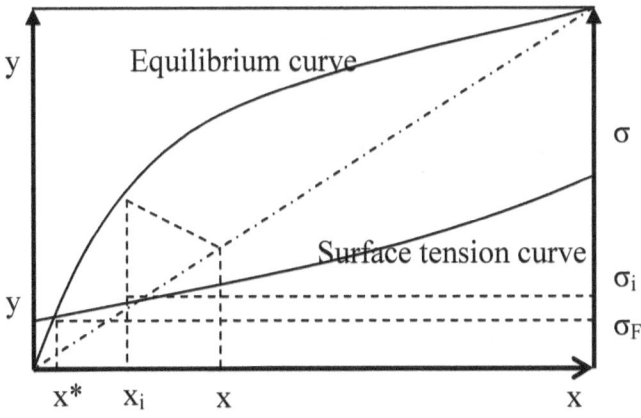

3.3.4.2: Temperature Effects which Lead to Marangoni Instabilities

From the Eotvos' formula

$$\sigma = \rho_M^{\frac{2}{3}} k_E (T_C - T - 6) \qquad (3.89)$$

where ρ_M = molar density, kmol/m³
 k_E = Eotvos constant (= 2.1 x 10⁻⁷ J/K.kmol⁻²/³ for normal liquids)
 T_C = critical temperature, deg. K
 σ = surface tension, N/m

Equation (3.89) shows that σ decreases with increase in temperature. In equipment with supported interfacial area, this decrease makes the system behave as a weakly surface tension negative system.

3.3.4.3: Influence of the Marangoni Effect on Mass Transfer Coefficients

The Marangoni effect can be observed either as roll cells (ordered flow) or as eruptions (disordered disturbances) depending on the physical state of the mass transfer system. These effects influence the mass transfer in the system differently.

a. <u>Roll Cells</u>

Roll cells occur in quiescent liquid as a result of surface tension differences between the bulk and surface of the liquid. The surface layer sinks, by gravity, while bulk liquid rises up, in distinct cells, as a result of having a lower surface tension than the surface. The direction of mass transfer is from the bulk to the surface.

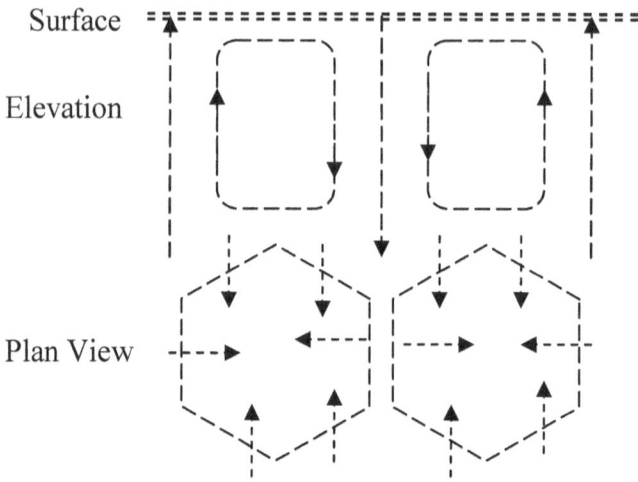

These movements lead to increases in surface renewal and surface tension instability controls mass transfer.

b. <u>Disordered Disturbances</u>

These occur when a packet of material rises, suddenly, from the bulk of the liquid to the surface. This may be illustrated as shown below.

··············	Initially, surface is quiescent
········■ ■ ■ ■ ■···········	Instantaneous bulk packet movement leads to spreading at the surface

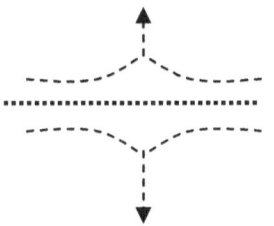 After introduction of packet, surface pressure leads to surface movements which result in jet or violent eruption

These eruptions increase mass transfer coefficients more than roll cell instability.

3.3.5: Surface Renewal

An important interface phenomenon, in mass transfer, is that of surface renewal. It is difficult to imagine, knowing how natural processes work, that an interface, in contact with a region high in the concentration of the mass being transferred, would not be exchanging that material with the adjacent region at the lower concentration of the material being transferred. If it did not pass on material that it received to the adjacent region of lower concentration, it would become saturated and mass transfer would stop as soon as all parts of the interface reach their equilibrium or steady state value. This is not observed to be the case in practice so that surface renewal is a reality in mass transfer through interfaces.

It becomes necessary then to determine how this renewal occurs and to derive quantitative relationships which make it possible to estimate their contribution or obstacle to mass transfer. It does not take much of an imagination to discern that the most appropriate analysis of surface renewal would have to be based on statistical methods.

3.3.5.1: The Cumulative Population Distribution Function

The cumulative distribution function, for age of a population, for example, is defined as

$$F(\theta) = \frac{\text{Population in age group } 0 \to \theta, \, p_0^\theta}{\text{Total Population, } P}$$

$$= \frac{p_0^\theta}{P} \qquad (3.90)$$

It follows that

$$dF(\theta) = \frac{\text{Incremental population in age group } \theta \to \theta + d\theta, \, p_\theta^{\theta+d\theta}}{\text{Total Population, } P}$$

$$= \frac{p_\theta^{\theta+d\theta}}{P} = \frac{dp_\theta}{P} = \left(\frac{dF(\theta)}{d\theta}\right) d\theta \qquad (3.91)$$

Graphically, we can see that the cumulative distribution curve is as shown below

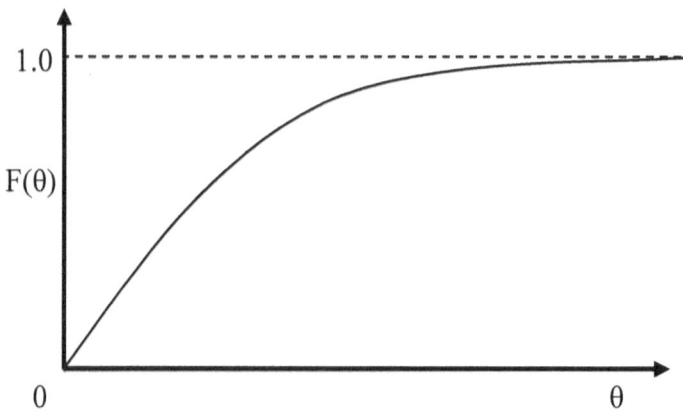

We can replace population by interfacial area in equation (3.91) to obtain that

$$\frac{dF(\theta)}{d\theta} = \frac{1}{A}\left(\frac{dA_\theta}{d\theta}\right) d\theta = \varphi(\theta) \qquad (3.92)$$

Note that $\dfrac{dF(\theta)}{d\theta}$ is the slope of the cumulative frequency distribution function.

$\varphi(\theta)$ is the frequency distribution function. It gives the probability of the area of an element being in a particular age group, θ. Graphically, the frequency distribution function looks like this

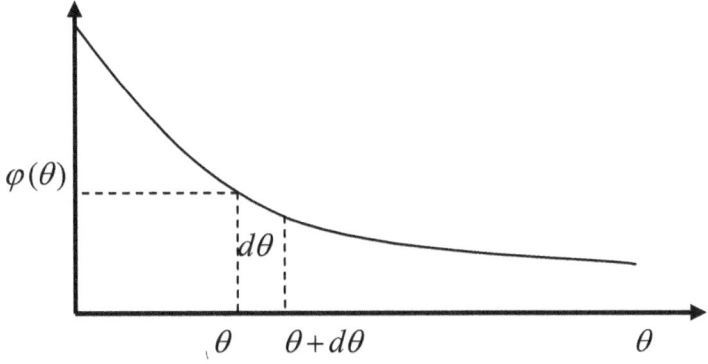

3.3.5.2: The Surface Age Distribution Function

It can be seen, from equation (3.92), that

$$\varphi(\theta) = \frac{dA_\theta}{A}$$

$$= \text{Fraction of surface area in age group } \theta \text{ to } d\theta \quad (3.93)$$

This is seen to be a more convenient definition with the added advantage that

$$\int_0^{\theta_{max}} \varphi(\theta) d\theta = \int_0^A \frac{dA_\theta}{A} = 1 \quad (3.94)$$

3.3.5.3: Average Mass Transfer Rate across an Interface, N_{Ai}

If $dW(\theta)$ is the rate of absorption into area in the age group θ to $\theta + d\theta$,

$$dW(\theta) = N_{A_i}(\theta).dA(\theta) \quad (3.95)$$

That is

$$N_{A_i} = \frac{W}{A} = \frac{1}{A}\int_0^\infty dW(\theta) = \frac{1}{A}\int_0^\infty N_{A_i}(\theta) dA(\theta)$$

$$= \int N_{A_i}(\theta)\varphi(\theta) d(\theta) \quad (3.96)$$

3.3.5.4: General Form of the Surface Age Distribution Function

We found from equations (3.94) and (3.96) that $\int_0^{\theta_{max}} \varphi(\theta) d\theta = 1$ and that $N_{A_i} = \int N_{A_i}(\theta) \varphi(\theta) d(\theta)$. It is clear that we need to know the form of $\varphi(\theta)$ in order to be able to evaluate equation (3.96).

Surface area in age group θ to $\theta + d\theta$
$\quad = $ *surface area in age group $\theta - d\theta$ to θ*
$\quad - $ *surface area renewed in time $d\theta$*

That is $\quad dA_\theta = dA_{\theta - d\theta} - s(dA_{\theta - d\theta}) d\theta \qquad (3.97)$

where $s = $ *fractional rate of surface renewal* $= \dfrac{\Delta A/A}{\Delta t} \qquad (3.98)$

Dividing equation (3.97) throughout by A and combining the result with equation (3.93) we get

$$\varphi(\theta) d\theta = \varphi(\theta - d\theta) d\theta - s[\varphi(\theta - d\theta) d\theta] d\theta \qquad (3.99)$$

When we simplify equation (3.99) by dividing by $d\theta$, remembering the fundamental definition of the differential in calculus, we get that

$$\varphi(\theta) = \varphi(\theta) - \dfrac{d\varphi(\theta)}{d\theta} d\theta - s.[\varphi(\theta) - d\varphi(\theta)] d\theta$$

So that $d\varphi(\theta) = -s\varphi(\theta) d\theta \quad$ or $\quad \dfrac{d\varphi(\theta)}{\varphi(\theta)} = -s d\theta \qquad (3.100)$

When equation (3.100) is integrated, we get that
$$\ln \varphi(\theta) = -\int s d\theta + cons \tan t$$

or $\quad \varphi(\theta) = cons\tan t . e^{-\int s d\theta} \qquad (3.101)$

$\varphi(\theta)$ will be different for different models of surface renewal but the value of the constant in equation (3.101) can always be evaluated using equation (3.94).

CHAPTER THREE: INTERFACIAL PHENOMENA

3.3.5.5: Models of Surface Renewal

3.3.5.5.1: The Higbie Model – The Regular Surface Renewal Model

Assumptions
  i. Time of contact, θ_C, is constant. That is, all surface elements are exposed to gas for the same duration of time
  ii. For $0 < \theta < \theta_C$, no element is destroyed unless $\theta = \theta_C$ i.e. $s = 0$
  iii. When $\theta = \theta_C$, surface is renewed instantaneously i.e. $s = \infty$

Thus, from the second assumption above, for $0 < \theta < \theta_C$, $s = 0$, and equation (3.101) becomes

$$\varphi(\theta) = cons\tan t . e^{-(0)} = cons\tan t \qquad (3.102)$$

From equations (3.94) and (3.102)

$$cons\tan t . \int_0^{\theta_C} d\theta = 1 \quad or \quad cons\tan t = \frac{1}{\theta_C} \qquad (3.103)$$

Substituting the value of the constant from equation (3.103) into equation (3.101) we get the general expression for the Higbie model as

$$\varphi(\theta) = \frac{1}{\theta_C} e^{-\int_0^{\theta_C} s\, d\theta} \qquad (3.104)$$

From the third assumption, when $\theta = \theta_C$, $s = \infty$ and equation (3.104) becomes

$$\varphi(\theta) = 0 \qquad (3.105)$$

Equations (3.102), (3.103) and (3.105) show that the Higbie model of surface renewal can be represented, graphically, as shown below.

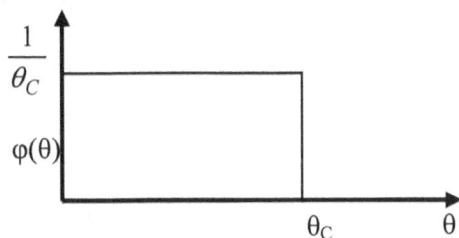

a. Applying the Higbie Model to the Penetration Theory

$$N_{A_i} = \int_0^{\theta_C} N_{A_i}(\theta)\varphi(\theta)d\theta = \frac{1}{\theta_C}\sqrt{\frac{D}{\pi}}.(C_{A_I} - C_{A_0})\int_0^{\theta_C}\frac{d\theta}{\sqrt{\theta}}$$

$$= 2\sqrt{\frac{D}{\pi\theta_C}}(C_{A_I} - C_{A_0}) = 2N_{A_i}(\theta_C) \qquad (3.106)$$

Equation (3.106) is identical with those for the unsteady state diffusion into a stagnant medium or into a falling film exposed to diffusion for a time θ_C. The mass transfer coefficient is

$$K = \frac{N_{A_i}}{C_{A_I} - C_{A_0}} = 2\sqrt{\frac{D}{\pi\theta_C}} \propto D^{0.5} \qquad (3.107)$$

b. Applying the Higbie Model to the Laplace Transform Solution of the film penetration theory, for $D\theta/L^2 < 1/\pi$,

$$N_{A_i} = 2.\sqrt{\frac{D}{\pi\theta_C}}(C_{A_I} - C_{A_0})\left[1 + 2\sqrt{\pi}\sum_{n=1}^{\infty}\int erfc\left(\frac{nL}{\sqrt{D\theta_C}}\right)\right] \qquad (3.108)$$

And To The Separation Of Variables Solution, $D\theta/L^2 > 1/\pi$,

$$N_{A_i} = \frac{D}{L}(C_{A_I} - C_{A_0})\left[1 + \frac{2}{\pi^2}.\frac{L^2}{D\theta_C}.\sum_{n=1}^{\infty}\frac{\pi^2}{6} - e^{-\left(\frac{n^2\pi^2 D\theta_C}{L^2}\right)}\right] \qquad (3.109)$$

The mass transfer coefficient is seen to be related to D to a power between 0.5 and 1.0.

3.3.5.5.2: The Danckwert's Model – The Random Surface Renewal Model

Assumption:

Old and new elements have the same probability of renewal. That is: s is a constant and not a function of time.

Thus, from equation (3.101)

$$\varphi(\theta) = cons\tan t.e^{-\int s d\theta} = cons\tan t.e^{-s\theta} \qquad (3.110)$$

From equations (3.94) and (3.110)

CHAPTER THREE: INTERFACIAL PHENOMENA

$$\int_0^\infty \varphi(\theta)\,d\theta = 1 = cons\tan t \int_0^\infty e^{-s\theta}\,d\theta$$

$$= cons\tan t \left(-\frac{1}{s}\right).\left|e^{-s\theta}\right|_0^\infty = \frac{cons\tan t}{s}$$

That is: $cons\tan t = s$ and $\varphi(\theta) = se^{-s\theta}$ (3.111)

Graphically, equation (3.11) may be illustrated as follows

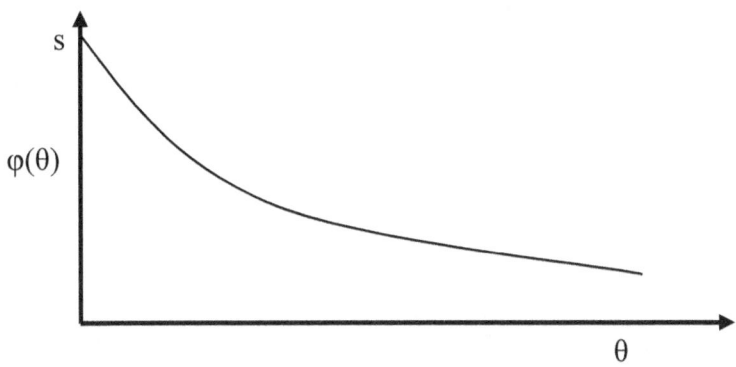

a. Applying The Danckwert's Random Surface Renewal Model To The Penetration Theory

Since $N_{A_I} = \sqrt{\dfrac{D}{\pi\theta}}(C_{A_I} - C_{A_0})$

$$N_{A_I} = \int_0^{\theta_c} N_{A_I}(\theta)\varphi(\theta)\,d\theta = \int_0^\infty \sqrt{\frac{D}{\pi\theta}}.(C_{A_I} - C_{A_0})se^{-s\theta}\,d\theta$$

$$= \sqrt{\frac{D}{\pi}}(C_{A_I} - C_{A_0})s \int_0^\infty e^{-s\theta}\frac{d\theta}{\sqrt{\theta}} = 2.\sqrt{\frac{D}{\pi}}(C_{A_I} - C_{A_0})s \int_0^\infty e^{-s\theta}\,d\sqrt{\theta}$$

$$= 2.\sqrt{\frac{D}{\pi}}(C_{A_I} - C_{A_0})\sqrt{s} \int_0^\infty e^{-s\theta}\,d\sqrt{s\theta}$$

$$= \sqrt{Ds}(C_{A_I} - C_{A_0})\frac{2}{\sqrt{\pi}} \int_0^\infty e^{-s\theta}\,d\sqrt{s\theta} = \sqrt{Ds}(C_{A_I} - C_{A_0}) \quad (3.112)$$

Equation (3.112) is true because

$$\frac{2}{\sqrt{\pi}} \int_0^\infty e^{-s\theta}\,d\sqrt{s\theta} = erfc(\infty) = 1 \quad (3.113)$$

The mass transfer coefficient, according to the Danckwert's model is then, from equation (3.113),

$$K = \frac{N_{A_i}}{C_{A_i} - C_{A_0}} = \sqrt{Ds} \propto D^{0.5} \qquad (3.114)$$

b. Applying The Danckwert's Random Surface Renewal Model To The Film Penetration Theory

The film penetration theory gave the molar flux, for $D\theta_C/L^2 < 1/\pi$,

$$N_{A_i} = 2\sqrt{\frac{D}{\pi\theta_C}}(C_{A_i} - C_{A_0})\left[1 + 2\sqrt{\pi}\sum_{n=1}^{\infty}\int erfc\left(\frac{nL}{D\theta_C}\right)\right] \qquad (3.108)$$ and

for $D\theta_C/L^2 > 1/\pi$,

$$N_{A_i} = \frac{D}{L}(C_{A_i} - C_{A_0})\left[1 + \frac{2}{\pi^2}\cdot\frac{L^2}{D\theta_C}\cdot\sum_{n=1}^{\infty}\frac{\pi^2}{6} - e^{-\left(\frac{n^2\pi^2 D\theta_C}{L^2}\right)}\right] \qquad (3.109)$$

Before the Danckwert's model is applied to these equations, the first thing one notices is that the part of the equations, with an infinity limit, is not really important because mass transfer to old surfaces is much less than that to young surfaces. Thus, that part of the equation with an upper limit of infinity is really negligible. The equations reduce, therefore, to either the penetration theory or the Whitman film theory models. Dr. Perlmutter (1961) also found that the residence time distribution function obtained with the film penetration model using the Danckwert's model of $\varphi(\theta)$ was not logical.

3.3.5.5.3: The Multiple Capacitance Model

<u>Relationships between Residence Time, and Surface Age, Distribution Functions</u>

Let F(t) = cumulative surface residence time distribution Function

f(t) = surface residence time frequency distribution function

Then

$$f(t) = \frac{dF(t)}{dt} \quad or \quad F(t) = \int_0^t f(t)dt \qquad (3.115)$$

with F(0) = 0 and F(∞) = 1.

CHAPTER THREE: INTERFACIAL PHENOMENA

If Q_S is the rate of supply of fresh surface to the interface of area S, then

$$dS(\theta) = Q_S d\theta - F(\theta)Q_S d\theta = Q_S[1-F(\theta)]d\theta \quad (3.116)$$

But

$$Q_S = sS \quad and \quad \tau = \frac{S}{Q_S} = \frac{1}{s} \quad (3.117)$$

so that

$$\varphi(\theta) = s[1-F(\theta)] = \frac{1}{\tau}[1-F(\theta)] \quad (3.118)$$

Thus, for a given $\varphi(\theta)$ f(t)

$$F(\theta) = 1 - \tau\varphi(\theta)$$

$$F(t) = \int_0^t f(t)\,dt$$

$$F(t) = F(\theta)$$

$$F(\theta) = F(t)$$

$$f(t) = \frac{dF(t)}{dt}$$

$$\varphi(\theta) = \frac{1}{\tau}[1-F(\theta)]$$

When these are applied to the various surface renewal models, we get that

a. <u>For the Higbie Model</u>

$$\varphi(\theta) = \frac{1}{\theta_C} = \frac{1}{\tau} \qquad F(t) = 0 \qquad 0 < t < \tau$$

$$\varphi(\theta) = 0 \qquad F(t) = 1 \qquad t > \tau$$

The Higbie model depicts the surface phenomenon as a plug flow surface reactor with the cumulative distribution function below

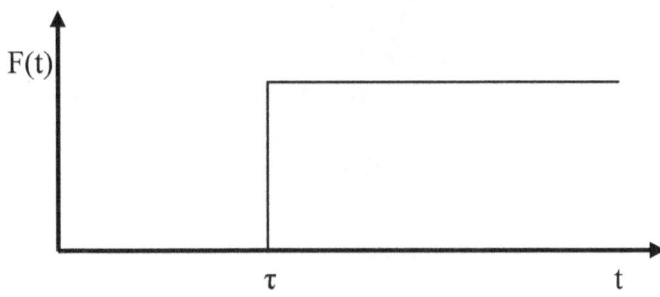

b. <u>For the Danckwert's Model</u>

$$\varphi(\theta) = \frac{1}{\tau} e^{-\frac{\theta}{\tau}} \qquad\qquad F(\theta) = 1 - e^{-\frac{\theta}{\tau}}$$

$$F(t) = 1 - e^{-\frac{t}{\tau}} \qquad\qquad f(t) = \frac{1}{\tau} e^{-\frac{t}{\tau}}$$

The surface behaves likes a single stirred tank surface reactor with the frequency distribution function

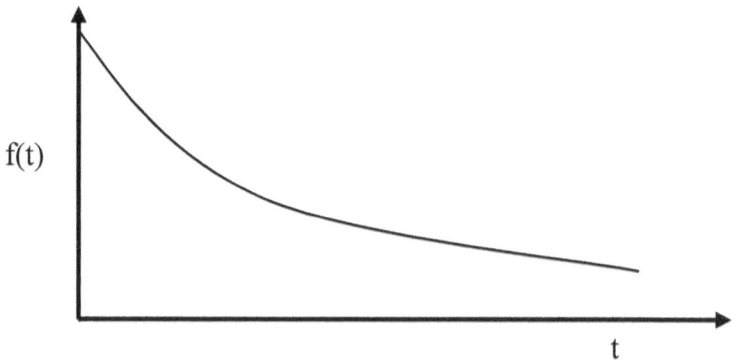

c. <u>The Multiple Capacitance Model</u>

This model assumes that the degree of mixing in the surface zone is specified by a number of stirred two-dimensional surface vessels connected in series. That is

$$f(t) = \frac{n^n t^{n-1}}{(n-1)!\tau^n} e^{-\frac{nt}{\tau}} \qquad (3.119)$$

and

$$F(t) = \int_0^\infty f(t)\,dt = \int_0^\infty \frac{n^n t^{n-1}}{(n-1)!\tau^n} e^{-\frac{nt}{\tau}} dt \qquad (3.120)$$

If n = ∞, f(t) = 0 or F(t) = constant, that is the Higbie model while if n = 1, it represents the Danckwert's model. If n = 2, it becomes the double capacitance model.

d. <u>The Double Capacitance Model</u>

For n = 2 and $\tau_1 = \tau_2 = \tau$

$$f(t) = \frac{4t}{\tau^2} e^{-\frac{2t}{\tau}} \tag{3.121}$$

From equations (3.115) and (3.121) we obtain that

$$F(t) = 1 - e^{-\frac{2t}{\tau}} - \frac{2t}{\tau} e^{-\frac{2t}{\tau}} \tag{3.122}$$

$$F(\theta) = 1 - e^{-\frac{2\theta}{\tau}} - \frac{2\theta}{\tau} e^{-\frac{2\theta}{\tau}} \tag{3.123}$$

$$\varphi(\theta) = \frac{1}{\tau}\left(1 + \frac{2\theta}{\tau}\right) e^{-\frac{2\theta}{\tau}} \tag{3.124}$$

e: Applying the Double Capacitance Model to the Penetration Theory

From equation (3.124), with $s = 1/\tau$ and

$$N_{A_i}(\theta) = \sqrt{\frac{D_{AB}}{\pi\theta}}(C_{A_I} - C_{A_0}) \text{ and } N_{A_i} = \int_0^{\theta_{max}} N_{A_i}(\theta)\varphi(\theta)d\theta$$

$$N_{A_i} = \int_0^\infty N_{A_i}(\theta)\varphi(\theta)d\theta = \sqrt{\frac{D}{\pi}}.(C_{A_I} - C_{A_0})s\int_0^\infty \frac{(1 + 2s\theta)}{\sqrt{\theta}} e^{-2s\theta} d\theta$$

$$= \frac{3.\sqrt{2}}{4}\sqrt{D_{AB}.s}(C_{A_I} - C_{A_0}) \tag{3.125}$$

The mass transfer coefficient becomes

$$K = \frac{N_{A_i}}{C_{A_I} - C_{A_0}} = \frac{3.\sqrt{2}}{4}\sqrt{D_{AB} s} = 1.06\sqrt{D_{AB} s} \tag{3.126}$$

Note that when the Higbie model of surface renewal (equivalent to n = ∞) is applied to the penetration theory,

$$N_{A_i} = 2\sqrt{\frac{D}{\pi\theta_c}}(C_{A_I} - C_{A_0}) = 1.128\sqrt{\frac{D}{\tau}}(C_{A_I} - C_{A_0}) \tag{3.106a}$$

The multiple capacitance model is seen to be the Higbie model, for n = ∞, the Danckwert's model for n = 1, and to only introduce correction factors (1.06) of order of experimental errors when n = 2. It has, in addition, too many parameters and is, therefore, not really useful.

3.3.5.5.4: The Surface Rejuvenation (Random Eddy) Model of the Penetration Theory

Assumptions of the Model

Large eddies bring material from the bulk, which is of uniform concentration, up to a distance, h, from the interface. This approach distance, h, varies randomly between 0 and ∞. Similarly, the residence times of eddies, in the vicinity of the interface, are subject to random variations between 0 and ∞, during which mass is transferred into the eddies by molecular diffusion. The significant depth of penetration is always smaller than the thickness of the eddies and there is a constant concentration of the solute at the interface. The solute concentration is small enough for zero net flux conditions to apply. The schematics, below, illustrate the physical situation as well as the concentration profiles with depth of penetration.

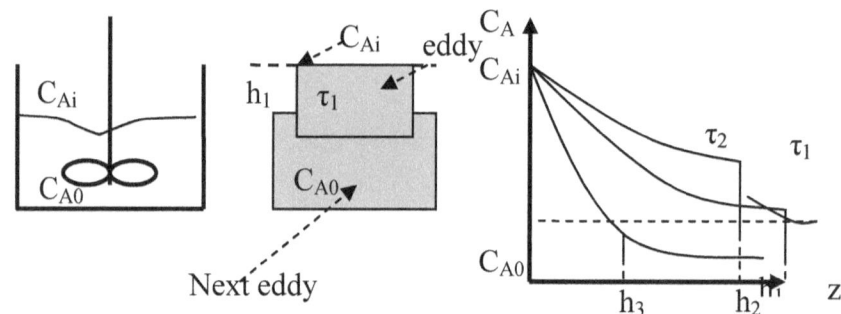

Mathematical Formulation

The equation to be solved is

$$\frac{\partial C_A}{\partial \theta} = D\left[\frac{\partial^2 C_A}{\partial z^2}\right] \qquad (2.15)$$

with the following boundary conditions:

1	$\theta = 0$	$z \geq 0$	C_A	=	C_{A0}
2	$\theta \geq 0$	$z = 0$	C_A	=	C_{Ai}	(2.31)
3	$\theta \geq 0$	$z = \infty$	C_A	=	C_{A0}

CHAPTER THREE: INTERFACIAL PHENOMENA

This had the solution

$$C_A = C_{A_0} + (C_{A_i} - C_{A_0}) \, erfc \frac{z}{2\sqrt{\theta D_{AB}}} \qquad (2.48)$$

If we assume that at $t = 0$, $h = 0$, then at $t = \tau_1$, a fresh eddy arrives up to a distance $z = h_1$ from the interface. The new boundary and initial conditions for this state of affairs become

$$\begin{array}{llll}
1 & t \geq 0 & z = \infty & C_A = C_{A0} \\
2 & t > 0 & z = 0 & C_A = C_{Ai} \\
3 & \theta_1 = t - \tau_1 & z > 0 & C_A = C_A(\theta, z)
\end{array} \qquad (3.127)$$

where $C_A(\theta, z)$ is given by equation (2.48) above for $0 < z < h_1$ and by $C_A = C_{A0}$ for $h_1 < z < \infty$. The general boundary and initial conditions, following from (3.127), are

$$\begin{array}{llll}
1 & t \geq 0 & z = \infty & C_A = C_{A0} \\
2 & t > 0 & z = 0 & C_A = C_{Ai} \\
3 & \theta_j = t - \sum_{i=1}^{j} \tau_i & 0 < z < \infty & C_A = f(z)
\end{array} \qquad (3.128)$$

Equation (2.15) can be solved with the new boundary and initial conditions (3.128) (See Carslaw and Jaeger, Conduction of Heat in Solids for solutions with boundary and initial conditions (3.127)).

Average Molar Flux at the Interface

If $W(t)$ is the amount absorbed per unit interfacial area up to the time, t,

$$W(t) = \int_0^t N_{A_i}(t) \, dt = D_{AB} \int_0^t \left(\frac{\partial C_A}{\partial z} \right)_{z=0} dt \qquad (3.129)$$

$$N_{A_i} = \frac{W(t)}{t} = \frac{1}{t} \int_0^t N_{A_i}(t) \, dt \qquad (3.130)$$

Up to $t = \tau_1$,

$$W(\tau_1) = 2 \sqrt{\frac{D_{AB} \tau_1}{\pi}} \cdot (C_{A_i} - C_{A_0}) \qquad (3.131)$$

$$N_{A_i} = 2 \cdot \sqrt{\frac{D_{AB}}{\pi \tau_1}} \cdot (C_{A_i} - C_{A_0}) \qquad (3.132)$$

In general,

$$W(t) = \sum_{i=1}^{j}\left[W(\tau_i) + W(\theta_j)\right] \qquad (3.133)$$

$$\text{For } t = \sum_{i=1}^{j} \tau_i, \quad W(t) = \sum_{i=1}^{j} W(\tau_i) \qquad (3.134)$$

$$\text{and } N_{A_i} = \frac{\sum_{i=1}^{j} W(\tau_i)}{\sum_{i=1}^{j} \tau_i} \qquad (3.135)$$

Calculation Procedure

1. Select distribution functions of residence times and approach distances of eddies
2. Generate a sequence of random time intervals, τ_1, τ_2, τ_3.... And a corresponding sequence of random approach distances, $h_1, h_2, h_3, \ldots$
3. Start calculations by setting $h_0 = 0$ and $f(z) = C_{A0}$. Calculate $W(\tau_1)$ and N_{Ai} at $t = \tau_1$. The starting condition for the second round of calculations is that $C_A = f(\tau_1, z)$ for $z < h_1$ and $C_A = C_{A0}$ for $z > h_1$. Continue calculations until there is no further significant change in N_{Ai}.

REFERENCES

1. Class Notes, Imperial College of Science & Technology, London, 1966
2. Danckwerts P. V., Ind. Eng. Chem.; 1951, 43, 1460
3. Harriott P., Chem. Engrng Science; 1962, 17,149
4. Higbie R., Trans. Am. Inst. Chem. Engrsd; 1935, 31, 365
5. Perlmutter D. D., Chem. Engrng Science; 1961, 16, 287

CHAPTER FOUR
MASS TRANSFER WITHOUT CHEMICAL REACTION

4.0: Introduction

The previous chapters dealt with the basic principles of mass transfer by molecular and eddy diffusion and also as a function, or not, of time. Conditions at the interface, through which mass transfer occurs, were also, examined. The importance of combining the various properties and phenomena of the bulk and the interface in order to obtain accurate estimates of the mass transferred and its rates was, also, illustrated. This chapter will attempt to present the total picture, both mathematically and otherwise, of the relationships between phase and interface concentrations to the rates of mass transfer in various typical mass transfer operations and processes.

In spite of all the work done to explore the time and surface property dependencies in mass transfer, most commercial analysis of mass transfer are still based on the Whitman two film theory. The only exceptions are where the physical evidence suggests its significant inapplicability to the particular situation or that significant improvement in mass transfer efficiencies are obtained by the use of the more complicated time or surface property dependent analyses.

4.1: Mass Transfer of Component A in Two Phases

Consider the mass transfer of, say, component A between two phases labelled I and II through an interface. Let C_{AIIb}, C_{AIIi} be the concentration of component A in the bulk and interface side of phase II, respectively, while C_{AIb} and C_{AIi} represent tyhe concentration in the bulk and inteface side of phase I. The physical situation is illustrated in the schematic diagram below. The concentration profile shown is that due to the Whitman two film theory.

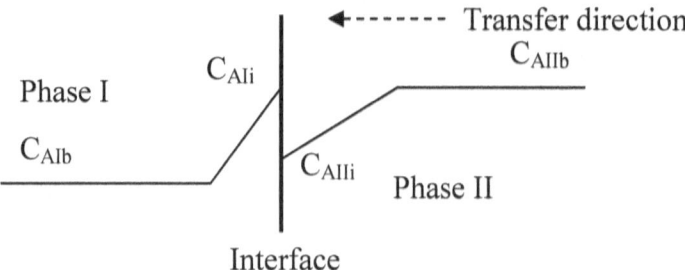

Interface

Recall that the phase rule states that the sum of the number of degrees of freedom, F, and the number of phases, P, in a PVT system, is equal to the number of components, C, in the system plus 2. That is

$$F + P = C + 2 \qquad (4.1)$$

For a two phase system with three components, the degree of freedom would be $3 + 2 - 2 = 3$. Similarly, for a three phase system with two components, the degree of freedom would be $2 + 2 - 3 = 1$.

The typical system we will use in our study is a two phase, three component system. Two phases because we encounter gas/liquid, liquid/liquid and solid/liquid phase combinations more often in commercial mass transfer operations than in more specialist applications, where there can be any number of phases. Three components because the phase from which mass is transfered can, minimally, be the pure solute or solute in a solvent while the phase to which mass is being transferred can be pure but different solvent. Even when there are more than three components in the system, the analysis may still be made as if it was a three component system by selecting a key diffusing component and a key solvent while lumping the properties of the other components together.

Thus, with two phases and three components we have three degrees of freedom, namely, temperature, pressure and composition. It is usual and economically attractive to operate at constant temperature and pressure, unless the particular situation demands otherwise. This leaves us with one degree of freedom, composition.

That means that if the composition is fixed in one phase, the composition in the other phase is fixed automatically. This is known as the equilibrium condition. This can be illustrated as shown below, for temperature and pressure constant.

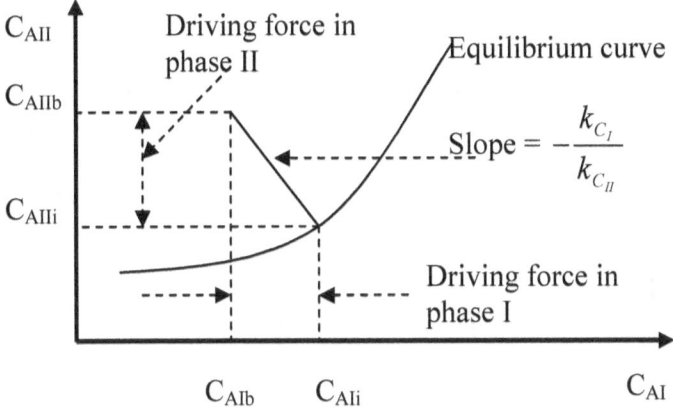

Note that this curve is valid only for the constant temperature and pressure on which it is generated. At other temperatures and pressures, different curves, though, essentially, of the same shape, will be generated. Note, also, that, although, C_{AIi} and C_{AIIi} are coincident on the equilibrium curve, their numerical valoues are not the same, as can be seen from the C_{AI} and C_{AII} axes. Remember, also, that we are dealing with the Whitman film mass transfer and are, therefore, referring to the films in phases I and II.

Since

$$N_{A_I} = -k_{C_I}(C_{I_b} - C_{I_i}) \quad (4.2)$$

$$N_{A_{II}} = k_{C_{II}}(C_{II_i} - C_{II_b}) \quad (4.3)$$

where k_{CI} and k_{CII} are the mass transfer coefficients in the films formed in phase I and II, respectively. Under steady state conditions, $N_{AI} = N_{AII}$ so that

$$\frac{C_{II_b} - C_{II_i}}{C_{I_b} - C_{I_i}} = -\frac{k_{C_I}}{k_{C_{II}}} \quad (4.4)$$

This shows that the ratio of the film mass transfer coefficients corresponds to the slope of the line joining the bulk and interface concentrations. It is useful in determining interface concentrations when the bulk concentration and the equilibrium curve are known.

4.1.1: Gas Phase and Liquid Phase in Equilibrium

When the phases involved are gas and liquid, the principles discussed above still apply but the concentration in each phase is represented differently. The concentration in the gas phase, our previous phase II, is represented by partial pressure, p, or by mole fraction, y, while the concentration in the liquid phase, our previous phase I, is represented by liquid concentration, C, or mole fraction, x. If, for example, we are dealing with absorption/desorption, the curves, at constant temperature and pressure, can be represented, schematically, as follows

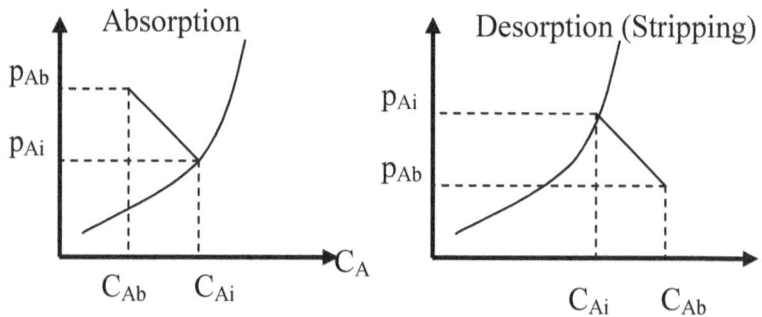

p_A is partial pressure of A in the gas phase

C_A is concentration of A in the liquid phase

It can be seen, from equation (4.4), that

$$\frac{p_{A_b} - p_{A_i}}{C_{A_b} - C_{A_i}} = -\frac{k_C}{k_p} \qquad (4.5)$$

That is, partial pressure driving force is proportional to concentration driving force.

4.1.2: Overall Mass Transfer Coefficient, K

There are many cases, in practice, where the equilibrium curve is a straight line or may be considered straight in a region of interest. In such cases, the equilibrium curve may be represented as

$$p_{A_i} = mC_{A_i} + b \qquad (4.6)$$

where m is the slope of the equilibrium line and b is a constant. If k_C/k_p is constant in this region of interest, we can represent the equilibrium line, the bulk and interface concentrations and partial pressures in the manner shown below.

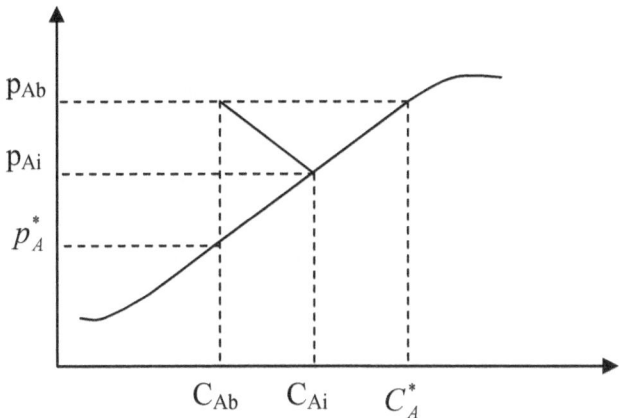

The figure above shows, also, that

$$p_A^* = mC_{A_b} + b \qquad (4.7)$$

From equations (4.6)

$$p_{A_b} - p_{A_i} = p_{A_b} - mC_{A_i} - b$$

or $$p_{A_b} = (p_{A_b} - p_{A_i}) + mC_{A_i} + b \qquad (4.8)$$

Similarly, from equation (4.7)

$$p_{A_b} - p_A^* = p_{A_b} - mC_{A_b} - b$$
$$= (p_{A_b} - p_{A_i}) + mC_{A_i} + b - mC_{A_b} - b$$
$$= (p_{A_b} - p_{A_i}) - m(C_{A_b} - C_{A_i})$$
$$\propto (p_{A_b} - p_{A_i}) \qquad (4.9)$$

Thus, from equation (4.3)

$$N_A = k_p(p_{A_b} - p_{A_i}) = K_G(p_{A_b} - p_A^*) \quad (4.10)$$

K_G is an overall mass transfer coefficient, expressed in terms of partial pressure driving force in the gas phase. It accounts for diffusion in both the gas and liquid films postulated by the Whitman film theory. A similar expression can be derived for the overall mass transfer coefficient, K_L, based on the concentration driving force in the liquid phase. Similarly

$$N_A = k_C(C_{A_b} - C_{A_i}) = K_L(C_A^* - C_{A_b}) \quad (4.11)$$

4.1.3: Relationships between Overall and Individual Film Mass Transfer Coefficients

The problem is, usually, to determine the molar flux or the mass transfer coefficients. For gas phase based mass transfer coefficients, equation (4.10) is

$$N_A = k_p(p_{A_b} - p_{A_i}) = K_G(p_{A_b} - p_A^*) \quad (4.10)$$

From equation (4.9)

$$p_{A_b} - p_A^* = (p_{A_b} - p_{A_i}) - m(C_{A_b} - C_{A_i}) \quad (4.9)$$

From equation (4.5)

$$C_{A_b} - C_{A_i} = -\frac{k_p}{k_C}(p_{A_b} - p_{A_i}) \quad (4.5a)$$

Substituting equation (4.5a) for the liquid concentration term in equation (4.9)

$$p_{A_b} - p_A^* = (p_{A_b} - p_{A_i}) + m\frac{k_p}{k_C}(p_{A_b} - p_{A_i})$$

$$= (p_{A_b} - p_{A_i})\left(1 + m\frac{k_p}{k_C}\right) \quad (4.12)$$

But from equation (4.10)

$$p_{A_b} - p_A^* = \frac{N_A}{K_G} \quad (4.10a)$$

$$p_{A_b} - p_{A_i} = \frac{N_A}{k_p} \quad (4.10b)$$

Substituting equations (4.10a) and (4.10b) in equation (4.12) and simplifying

$$\frac{N_A}{K_G} = \frac{N_A}{k_p}\left(1 + m\frac{k_p}{k_C}\right) \quad \text{or} \quad \frac{1}{K_G} = \frac{1}{k_p} + \frac{m}{k_C} \quad (4.13)$$

Similarly, for liquid phase based mass transfer coefficients, the following expression can be derived

$$\frac{1}{K_L} = \frac{1}{mk_p} + \frac{1}{k_C} \quad (4.14)$$

CHAPTER FOUR: MASS TRANSFER WITHOUT CHEMICAL REACTION

4.1.3.1: The Resistance Concept of Mass Transfer Coefficients

If we recall the fundamental concepts associated with transport processes, we will remember that

$$Flux = conductance \times driving\ force \qquad (1.35)$$

$$= \frac{Driving\ Force}{Resistance} \qquad (1.36)$$

Since the driving force for mass transfer is the partial pressure or concentration difference, it follows from any of the equations (4.10), (4.10a) or (4.10b) that the inverse of the mass transfer coefficient is the resistance to mass transfer. Thus the inverse of the overall mass transfer coefficient is the total resistance to mass transfer while the inverse of the gas or liquid film mass transfer coefficient is the resistance in the gas or liquid film, respectively. In terms of resistances, therefore, equation (4.13) states that, based on the gas film

$$Total\ Resistance = gas\ film\ resistance + m.(liquid\ film\ resistance) \qquad (4.15)$$

while equation (4.14) states that, based on the liquid film

$$Total\ Resistance = liquid\ film\ resistance + \frac{gas\ film\ resistance}{m} \qquad (4.16)$$

Equations (4.13) to (4.16) are extremely useful in commercial practice. For example, if m is small, m/k_C becomes negligible and

$$\frac{1}{K_G} = \frac{1}{k_p} \qquad (4.17)$$

That is, resistance to mass transfer is, almost, entirely, in the gas phase. Similarly, when m is large, $1/mk_p$ becomes negligible and

$$\frac{1}{K_L} = \frac{1}{k_C} \qquad (4.18)$$

That is, resistance to mass transfer is, almost, entirely, in the liquid phase.

4.2: Sizing Mass Transfer Process Equipment

Mass transfer process equipment operate in, essentially, two major modes, namely, as continuous processes and as batch processes. Continuous processes, in turn, can be staged (also called stage wise operations) or be by continuouis differential contact.

The basic calculation procedure is to obtain the mass balance across a representative element of the equipment. This mass balance should relate the mass or molar flux to the solute inlet and outlet concentrations as well as to an equilibrium relationship between concentrations of the solute in the two phases being processed. Evaluation of this material balance, as above, leads to an estimate of the height of tower or packing in wetted wall columns and packed towers, number of theoretical and actual plates in plate towers and liquid and gas flowrates with the concentration of components.

Problems encountered, which should not be problems really, are the ones of dimensional consistency, and customisation, of terms to suit particular industries.

Definitions of Mass Fraction, Mole Fraction and Mole Ratio

In the gas phase, the concentration of solute can be expressed as its partial pressure, p, its molar concentration, C, its mole fraction, y, and its mole ratio, Y. These are related to each other, for an ideal gas, as follows

$$p_A V = n_A RT \quad i.e \quad p_A = \frac{n_A}{V} RT = C_A RT \quad (4.19)$$

$$y_A = \frac{p_A}{p_A + p_B} = \frac{p_A}{P_T} = \text{mole fraction of } A \quad (4.20)$$

$$Y_A = \frac{y_A}{1 - y_A} = \frac{y_A}{y_B} = \frac{p_A}{p_B} = \text{mole ratio of } A \quad (4.21)$$

$$y_A = \frac{Y_A}{1 + Y_A} \quad \text{mole fraction in terms of mole ratio} \quad (4.22)$$

CHAPTER FOUR: MASS TRANSFER WITHOUT CHEMICAL REACTION

Note that, in practice, most gases of commercial importance are considered ideal gases if they are under a total pressure less than 20 atmospheres pressure.

In the liquid phase

$$x_A = \frac{n_A}{n_A + n_B} = \frac{C_A}{C_T} = \text{mole fraction of } A \quad (4.23)$$

$$X_A = \frac{x_A}{1 - x_A} = \frac{x_A}{x_B} = \frac{C_A}{C_B} = \text{mole ratio of } A \quad (4.24)$$

$$x_A = \frac{X_A}{1 + X_A} \quad \text{mole fraction in terms of mole ratio} \quad (4.25)$$

Further, to convert mole fraction of component, A, to mass fraction, multiply its mole fraction by the ratio of the mean molecular weight of the mixture to the molecular weight of component A. To convert mole ratio of component, A, to mass ratio, multiply its mole ratio by the molecular weight of component B divided by the molecular weight of component A.

4.2.1: Continuous Processes

Here, the gas and liquid phases are continuously in contact to facilitate mass transfer between them. Many types of equipment have been developed for this but the most common are the packed tower, several types of plate towers and the wetted wall column.

Operation can, also, be time dependent or steady state. Time dependent or unsteady state continuous operations usually occur at plant start up or shut down while steady state operations are preferred for consistency of product quality, system operation and maintenance.

Operation can be co-current or counter-current flow of the gas and liquid phases, which in liquid/liquid and solid/ liquid operations, are referred to as the lighter and heavier phases.

4.2.1.1: Differential Contact Operations

This can be the wetted wall column or, more commonly, the

packed tower.

Co- Current System

Phase bulk flow is said to be co-current because both phases flow into, through and from the system in the same direction. This may be represented, schematically, as follows

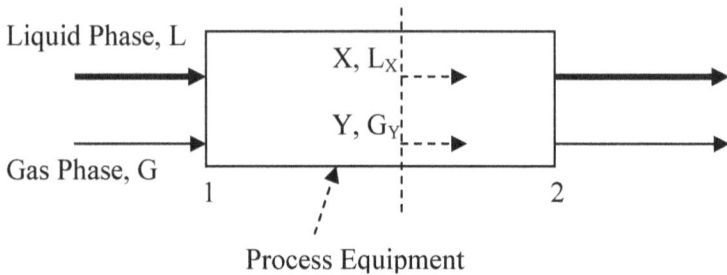

Process Equipment

Material Balance

Based on the law of conservation of mass, the general material balance, within a chosen boundary, is usually given as

Material In + Material Generated or Consumed = Material Out + Rate of Generation or Depletion. (4.26)

When there is no chemical reaction, no material is generated or consumed and equation (4.26) reduces to

Material In = Material Out + Rate of Accumulation (4.27)

When the operation is at steady state, the rate of accumulation is zero and equation (4.27) reduces to

Material In = Material Out . (4.28)

Let L and G are the bulk molal flowrates of the heavy and lighter phases (liquid and gas, in this example), respectively. Because of mass transfer, L_1 will not be the same as L_2 neither will G_1 be the same as G_2. In many situations where this is possible, it is more

CHAPTER FOUR: MASS TRANSFER WITHOUT CHEMICAL REACTION

convenient to define L and G so that they are free of component A. In such cases, L and G are constant throughout the process.

If Y_1, X_1, Y_2, X_2, are the mole ratios of component A in the gas and liquid phases, in entry and exit, respectively, an overall material balance, according to equation (4.28) gives

$$X_1 L_1 + Y_1 G_1 = X_2 L_2 + Y_2 G_2 \qquad (4.29)$$

or in terms of mole fraction

$$\frac{x_1}{1-x_1} L_1 + \frac{y_1}{1-y_1} G_1 = \frac{x_2}{1-x_2} L_2 + \frac{y_2}{1-y_2} G_2 \qquad (4.29a)$$

while a material balance between plane 1 and any plane within the process, where Y and X are the mole ratios, will give

$$X_1 L_1 + Y_1 G_1 = X L_X + Y G_Y \qquad (4.30)$$

or, in terms of mole fraction,

$$\frac{x_1}{1-x_1} L_1 + \frac{y_1}{1-y_1} G_1 = \frac{x}{1-x} L_X + \frac{y}{1-y} G_Y \qquad (4.30a)$$

Equations (4.29) and (4.30) represent the most general case of the steady state material balance in a co-current contactor. Equation (4.29) is useful for determining terminal compositions or flowrates if the other variables in the equation are known. Equation (4.30) is the so called operating line and gives the composition in both phases at any point in the system.

In practice, however, both equations (4.29) and (4.30) are difficult to use unless the variation of L and G along the contactor is known. A more convenient form of the equations is obtained by making L and G solute free, that is free of component A. When this is done, L and G become constant throughout the process and equations (4.29) and (4.30) become

$$X_1 L + Y_1 G = X_2 L + Y_2 G \quad \text{and} \quad L(X_1 - X_2) = G(Y_2 - Y_1)$$

or

$$\frac{L}{G} = \frac{Y_2 - Y_1}{X_1 - X_2} \qquad (4.31)$$

Equation (4.30) becomes

$$X_1 L + Y_1 G = X L + Y G$$

or $\quad Y = -\dfrac{L}{G}.X + \dfrac{L}{G} X_1 + Y_1 \quad$ the operating line $\qquad$ (4.32)

The operating line is a very important part of mass transfer process calculations.

In many cases, the liquid phase comes in free of component A. In such cases, $X_1 = 0$ and equation (4.32) reduces to a simpler straight line. In terms of mole fractions, equation (4.32) becomes

$$\frac{y}{1-y} = -\frac{L}{G}\cdot\frac{x}{1-x} + \frac{L}{G}X_1 + Y_1 \quad \text{the operating line} \quad (4.32a)$$

Equation (4.32a) can be reduced to the form

$$y = \frac{a-bx}{c-dx} \quad \text{where } a,b,c,d \text{ are constants} \quad (4.32b)$$

which is not, necessarily, a straight line.

Counter - Current System

Phase bulk flow is said to be counter-current because both phases flow into, through and from the system in opposite directions. This may be represented, schematically, as follows

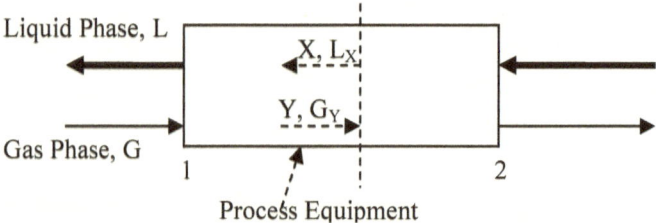

Material Balance

Based on the law of conservation of mass, the general material balance, within a chosen boundary, is usually given as, before, by

Material In + Material Generated or Consumed = Material Out + Rate of Accumulation (4.26)

When there is no chemical reaction, no material is generated or consumed and equation (4.26) reduces to

Material In = Material Out + Rate of Accumulation (4.27)

CHAPTER FOUR: MASS TRANSFER WITHOUT CHEMICAL REACTION

When the operation is at steady state, the rate of accumulation is zero and equation (4.27) reduces to

Material In = *Material Out* (4.28)

Let L and G are the bulk molal flowrates of the heavy and lighter phases (liquid and gas, in this example), respectively. Because of mass transfer, L_1 will not be the same as L_2 neither will G_1 be the same as G_2. In many situations where this is possible, it is more convenient to define L and G so that they are free of component A. In such cases, L and G are constant throughout the process.

If Y_1, X_1, Y_2, X_2, are the mole ratios of component A in the gas and liquid phases, in entry and exit, respectively, an overall material balance, according to equation (4.28) gives

$$X_2 L_2 + Y_1 G_1 = X_1 L_1 + Y_2 G_2 \quad (4.33)$$

or in terms of mole fraction

$$\frac{x_2}{1-x_2} L_2 + \frac{y_1}{1-y_1} G_1 = \frac{x_1}{1-x_1} L_1 + \frac{y_2}{1-y_2} G_2 \quad (4.33a)$$

while a material balance between plane 1 and any plane within the process, where Y and X are the mole ratios, will give

$$X L_X + Y_2 G_2 = X_2 L_2 + Y G_Y \quad (4.34)$$

or, in terms of mole fraction,

$$\frac{x}{1-x} L_X + \frac{y_2}{1-y_2} G_2 = \frac{x_2}{1-x_2} L_2 + \frac{y}{1-y} G_Y \quad (4.34a)$$

Equations (4.33) and (4.34), again, represent the most general case of the steady state material balance in a counter-current contactor. As in the co-current case, equation (4.33) is useful for determining terminal compositions or flowrates if the other variables in the equation are known. Equation (4.34) is the so called operating line and gives the composition in both phases at any point in the system.

As in the co-current case, equations (4.33) and (4.34) are difficult to use unless the variation of L and G along the contactor is known. A more convenient form of the equations is obtained by making L and G solute free, that is free of component A. When this is done, L and G become constant throughout the process and equations (4.33) and (4.34) become

$$X_2 L + Y_1 G = X_1 L + Y_2 G \quad \text{and} \quad L(X_2 - X_1) = G(Y_2 - Y_1)$$

or
$$\frac{L}{G} = \frac{Y_2 - Y_1}{X_2 - X_1} \qquad (4.35)$$

Equation (4.34) becomes
$$X L + Y_2 G = X_2 L + Y G$$

or $Y = \dfrac{L}{G} \cdot X - \dfrac{L}{G} X_2 + Y_2 \quad$ the operating line $\quad (4.36)$

In terms of mole fractions, equation (4.36) becomes

$$\frac{y}{1-y} = \frac{L}{G} \cdot \frac{x}{1-x} - \frac{L}{G} X_2 + Y_2 \quad \text{the operating line} \quad (4.36a)$$

Equation (4.36a) can be reduced to the form

$$y = \frac{a + bx}{c + dx} \quad \text{where } a, b, c, d \text{ are constants} \quad (4.36b)$$

which is not, necessarily, a straight line.

Referring to the schematic diagrams for co-current and counter-current operation, with the illustrated entry and exit points of both phases, it will be found that the Y-X diagrams will be different depending on whether the mass transfer process is one of absorption or desorption. These are shown below.

For absorption from the gas to the liquid phase, we have

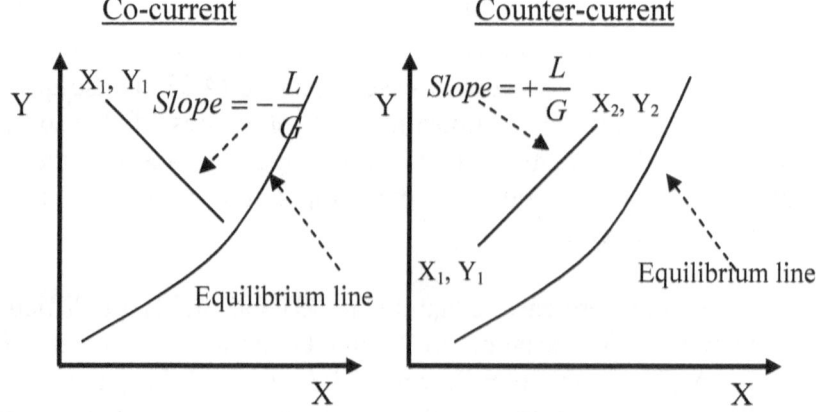

CHAPTER FOUR: MASS TRANSFER WITHOUT CHEMICAL REACTION

For desorption, from the gas to the liquid phase,

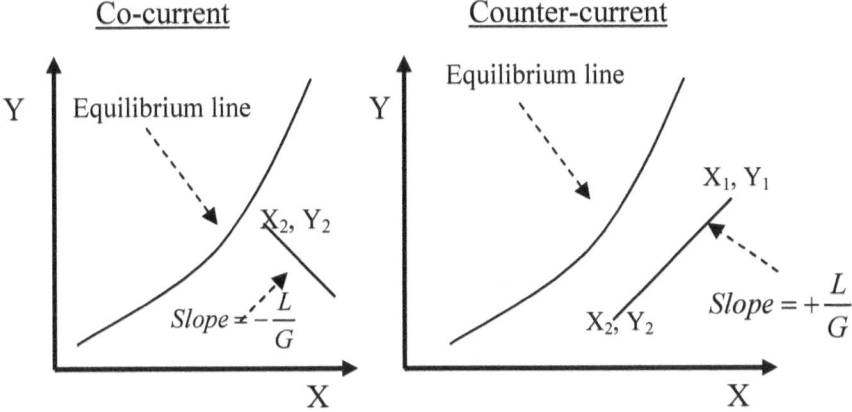

Estimating the Size/Capacity of the Differential Contactor

Consider a differential element of the mass transfer system, this time, a packed column. In order to relate molar flux and bulk flow stream concentrations to the size of the unit, we redefine the bulk flows in terms of the cross-sectional area of the tower and the surface area provided by the packing in terms of surface area per unit volume of the tower. That is, L and G are, now defined as moles per unit time per unit cross-sectional area of the tower. A schematic of the differential element, under consideration, is shown below.

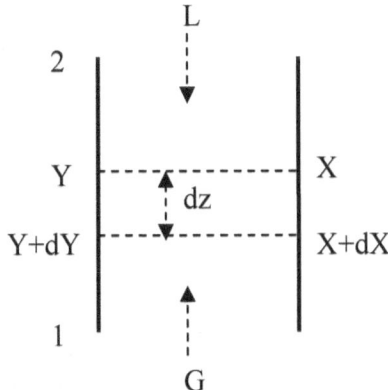

Hence, moles of A transferred per unit time over the height, dz, is

Moles transferred $= L \times A \times dX = G \times A \times dY$ (4.37)

where A is the cross-sectional area of the tower. The number of moles transferred is also

$$\text{Moles transferred} = N_A \times S \quad (4.38)$$

where S is effective surface area across which mass transfer occurs. This area is made up of the interfacial area provided by the packing which is defined as surface area per unit volume of the tower, a. Thus

$$S = a \times dV = a \times A \times dz \quad (4.39)$$

Equations (4.37), (4.38) and (4.39) become

$$N_A \cdot a \cdot A \cdot dz = L \cdot A \cdot dX = G \cdot A \cdot dY$$

or

$$N_A = \frac{L}{a} \cdot \frac{dX}{dz} = \frac{G}{a} \cdot \frac{dY}{dz} \quad (4.40)$$

From equations (4.10), (4.20) and (4.22),

$$N_A = k_y(y_{A_b} - y_{A_i}) = K_Y \left[\frac{Y_{A_b}}{Y_{A_b}+1} - \frac{Y_{A_i}}{Y_{A_i}+1} \right] \quad (4.41)$$

Substituting equation (4.41) into equation (4.40), we get that

$$\frac{G}{a}\frac{dY}{dz} = K_Y \left[\frac{Y_{A_b}}{Y_{A_b}+1} - \frac{Y_{A_i}}{Y_{A_i}+1} \right] \quad \text{from which we get that}$$

$$z = \int_0^z dz = \frac{G}{K_Y a} \int_{Y_{b_2}}^{Y_{b_1}} \frac{(1+Y_{A_b})(1+Y_{A_i})}{(Y_{A_b} - Y_{A_i})} dY \quad (4.42)$$

assuming that $K_Y a$ is constant throughout the column.

For dilute solutions, Y_{Ab} and Y_{Ai} are quite small so that their product, $Y_{Ab} \times Y_{Ai}$, is negligible. Thus, equation (4.42) reduces to

$$z = \int_0^z dz = \frac{G}{K_Y a} \int_{Y_{b_2}}^{Y_{b_1}} \frac{dY_{A_b}}{(Y_{A_b} - Y_{A_i})} \quad (4.43)$$

Similarly, using the X variable,

$$z = \int_0^z dz = \frac{L}{K_X a} \int_{X_{b_2}}^{X_{b_1}} \frac{dX_{A_b}}{(X_{A_i} - X_{A_b})} \quad (4.44)$$

Equations (4.42) and (4.43) are, easily, evaluated once the equilibrium curve, the volumetric surface area of the packing, a, and one or the other, and the ratio, of the film mass transfer coefficients are known.

CHAPTER FOUR: MASS TRANSFER WITHOUT CHEMICAL REACTION

The evaluation method is either by a graphical integration or by the use of numerical procedures such as the Simpson's rule or the Weddle's rule.

Transfer Unit Theory

As can be seen in the previous section, graphical procedures can be tedious and often make poor estimates of the performance of a stage, mainly because graphical procedures cannot be accurate to more than a few decimal places. New terms are defined, therefore, so that the height of the packing can be estiamted more accurately. One such definition is

Height of tower = height of transfer unit
$$\text{x number of transfer units} \quad (4.45)$$

If we recall from equation (4.43) that the height of the column was given as

$$z = \int_0^z dz = \frac{G}{k_Y a} \int_{y_{b2}}^{y_{b1}} \frac{d y_{A_b}}{(y_{A_b} - y_{A_i})} \quad (4.43)$$

then, it is not difficult to see that z can, also, be defined, according to equation (4.45) as

$$z = H_G \cdot N_G \quad (4.45a)$$

where $\quad H_G = \dfrac{G}{k_Y a} \quad$ and $\quad N_G = \displaystyle\int_{y_{b2}}^{y_{b1}} \dfrac{d y_{A_b}}{(y_{A_b} - y_{A_i})} \quad (4.46)$

H_G is defined as the height of transfer unit and N_G as the number of transfer units, often abreviated HTU and NTU respectively. A transfer unit is a unit of concentration change that is effected by unit driving force.

When overall mass transfer coefficients are used, the equivalent definitions are

$$H_{OG} = \frac{G}{K_Y a} \quad \text{and} \quad N_{OG} = \int_{y_{b2}}^{y_{b1}} \frac{d Y_{A_b}}{(Y_{A_b} - Y_A^*)} \quad (4.47)$$

where H_{OG} is the height of an overall mass transfer unit based on gas film resistance and N_{OG} is the overall number of transfer units based on gas film resistance.

The height of a transfer unit indicates a kind of general efficiency of the mass transfer process because, from equations (4.47) and the differential form of (4.43),

$$\frac{1}{H_{OG}} = \frac{\frac{dY_A}{dz}}{Y_A - Y_A^*} \qquad (4.47a)$$

expresses the separation achieved per unit driving force which is a measure of efficiency. The equivalent expressions for the height of transfer unit, H_L and H_{OL} and number of transfer units, N_L and N_{OL}, based on liquid film resistance are

$$H_L = \frac{L}{k_X a} \quad \text{and} \quad N_L = \int_{X_{b2}}^{X_{b1}} \frac{d x_{A_b}}{(x_{A_i} - x_{A_b})} \qquad (4.48)$$

$$H_{OL} = \frac{L}{K_X a} \quad \text{and} \quad N_{OL} = \int_{X_{b2}}^{X_{b1}} \frac{d X_{A_b}}{(X_A^* - X_{A_b})} \qquad (4.49)$$

Estimation of H_{OG} and N_{OG}

H_{OG} and H_{OL} can be obtained from experimental measurements of K_G and K_L while N_{OG} and N_{OL} can be obtained by a graphical or numerical integration of equations (4.47) and (4.49). An approach, used in heat transfer, which is also applicable to mass transfer, is the concept of the log mean driving force (log mean temperature difference in heat transfer). This is strictly valid in mass transfer only when the equilibrium line is a straight line.

Recall that Y* is obtained, for any point, (Y,X), as shown below

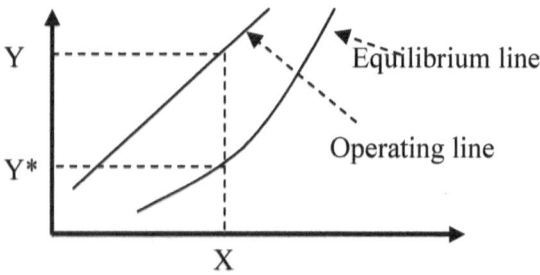

If the equilibrium curve is a straight line, its equation may be expressed as

$$Y_i = m X_i + b \qquad (4.50)$$

CHAPTER FOUR: MASS TRANSFER WITHOUT CHEMICAL REACTION

where m is the slope of the equilibrium curve and b is a constant. Since Y^*, also, lies on the equilibrium curve

$$Y^* = mX + b \tag{4.51}$$

and

$$Y - Y^* = Y - mX - b \tag{4.52}$$

The equation of the operating line, for a counter-current system is

$$Y = \frac{L}{G} \cdot X - \frac{L}{G} X_2 + Y_2 \tag{4.36}$$

from which

$$X = \frac{G}{L} \cdot Y + X_2 - \frac{G}{L} Y_2 \tag{4.53}$$

Substituting for X in equation (4.52)

$$Y - Y^* = Y - m\left[\frac{G}{L} \cdot Y + X_2 - \frac{G}{L} Y_2\right] - b$$

$$= Y\left(1 - \frac{mG}{L}\right) - m\left(X_2 - \frac{G}{L} Y_2\right) - b$$

$$= qY - r \tag{4.54}$$

where

$$q = \left(1 - \frac{mG}{L}\right) \quad \text{and} \quad r = m\left(X_2 - \frac{G}{L} Y_2\right) - b \tag{4.55}$$

Thus equation (4.47) becomes

$$N_{OG} = \int_{Y_{b2}}^{Y_{b1}} \frac{dY_{A_b}}{(Y_{A_b} - Y_A^*)} = \int_{Y_{A_2}}^{Y_{A_1}} \frac{dY_{A_b}}{(qY_{A_b} - r)} \tag{4.47}$$

$$= \left|\frac{1}{q} \ln(qY_{A_b} - r)\right|_{Y_{A_1}}^{Y_{A_2}} = \frac{1}{q} \ln\left(\frac{qY_{A_2} - r}{qY_{A_1} - r}\right)$$

$$= \frac{1}{q} \ln\left(\frac{Y_{A_2} - Y_{A_2}^*}{Y_{A_1} - Y_{A_1}^*}\right) \tag{4.56}$$

But from equation (4.54)

$$Y_{A_1} - Y_{A_1}^* = qY_{A_1} - r \tag{4.57}$$

$$Y_{A_2} - Y_{A_2}^* = qY_{A_2} - r \tag{4.58}$$

Subtracting equation (4.57) from equation (4.58)

$$(Y_{A_2} - Y_{A_2}^*) - (Y_{A_1} - Y_{A_1}^*) = q(Y_{A_2} - Y_{A_1})$$

$$\text{or} \quad q = \frac{(Y_{A_2} - Y_{A_2}^*) - (Y_{A_1} - Y_{A_1}^*)}{(Y_{A_2} - Y_{A_1})} \quad (4.58)$$

Substituting for q in equation (4.56)

$$N_{OG} = \frac{(Y_{A_2} - Y_{A_1})}{(Y_{A_2} - Y_{A_2}^*) - (Y_{A_1} - Y_{A_1}^*)} \ln\left(\frac{Y_{A_2} - Y_{A_2}^*}{Y_{A_1} - Y_{A_1}^*}\right)$$

$$= \frac{Y_{A_2} - Y_{A_1}}{[Y_A - Y_A^*]_{lm}} \quad (4.59)$$

where

$$[Y_A - Y_A^*]_{lm} = \frac{(Y_{A_2} - Y_{A_2}^*) - (Y_{A_1} - Y_{A_1}^*)}{\ln\left(\dfrac{Y_{A_2} - Y_{A_2}^*}{Y_{A_1} - Y_{A_1}^*}\right)} \quad (4.60)$$

When the equilibrium curve is not a straight line, a corrected form of the overall number of transfer units can be used. This is obtained by, first, calculating the arithmetic mean of the inlet and outlet concentration in the liquid stream. That is

$$X_m = \frac{X_{A_1} + X_{A_2}}{2} \quad (4.61)$$

This value is used in the Y-X diagram to locate Y_m and Y_m^*

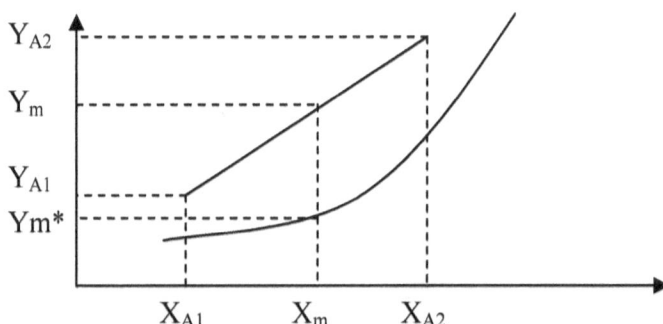

This enables us to obtain the arithmetic mean driving force, $(Y_A - Y_A^*)_{am}$ as

$$(Y_A - Y_A^*)_{am} = (Y_m - Y_m^*) \quad (4.62)$$

The overall number of transfer units, corrected for curvature of the equilibrium line, becomes

CHAPTER FOUR: MASS TRANSFER WITHOUT CHEMICAL REACTION

$$N_{OG} = \frac{Y_{A_2} - Y_{A_1}}{\left[Y_A - Y_A^*\right]_{am}} \cdot F \tag{4.63}$$

where F is a correcting factor given by

$$F = \frac{1}{2}\left[\frac{\ln \beta_{m_1}}{1 - \frac{1}{\beta_{m_1}}} + \frac{\ln \beta_{m_2}}{1 - \frac{1}{\beta_{m_2}}}\right] \tag{4.64}$$

where

$$\beta_{m_1} = \frac{(Y_A - Y_A^*)_{am}}{(Y_A - Y_A^*)_1} \quad \text{and} \quad \beta_{m_2} = \frac{(Y_A - Y_A^*)_{am}}{(Y_A - Y_A^*)_2} \tag{4.65}$$

Baker's Method

This method, which has its basis on transfer unit theory, is another procedure for using the overall number of transfer units to evaluate a system in which the equilibrium line is not a straight line. It assumes that the equilibrium line is straight over each transfer unit rather than over the whole tower operating conditions. It is a graphical procedure although it can be dapted to numerical solution. The steps are as follows:

1. draw the operating and equilibrium lines on the Y-X diagram
2. Construct a curve which joins the mid-points of the vertical intervals between the operating and equilibrium lines (Curve AB)
3. Starting from the liquid entry concentration point, C, draw a horizontal line, (CD), such that curve AB bisects this line.
4. from D, draw a vertical line to meet the operating line at E
5. continue from E as for C until end of operating point on the operating line.
6. Count the number and fraction, if any, of transfer units to get that
$$N_{OG} = number \; counted \; + \; fraction \tag{4.66}$$

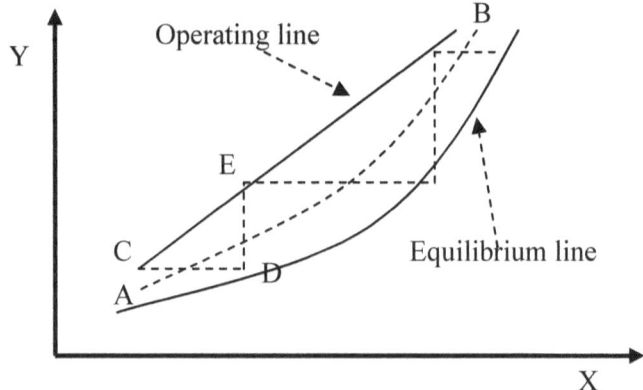

To find the fraction of N_{OG}, consider the portion of the Y-X diagram involved, as shown below.

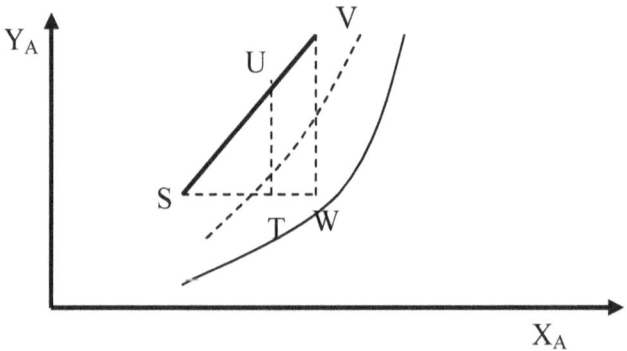

SU is the portion of the operating line in which the fraction of

N_{OG} resides. V is a point on a hypothetical continuation of the operating line to obtain a full transfer unit. The required fraction, ΔN_{OG} is given by $\Delta N_{OG} = \dfrac{TU}{WV}$ \hspace{1em} (4.67)

4.2.1.2: Stagewise Contact Operations

Stage wise operations aim to utilise the advantages and avoid the disadvantages of co-current and counter-current operations. Co-current operations have the advantage of approaching equilibrium conditions with the disadvantage that this requires very large, and hence, expensive equipment. Counter-current operations have the

CHAPTER FOUR: MASS TRANSFER WITHOUT CHEMICAL REACTION

advantage of smaller sized equipment but never reach equilibrium conditions. Hence, in stagewise operations, each stage is arranged to be co-current but the several stages, in the process, are arranged to operate counter-currently. A typical physical arrangement of a stagewise operation for gas absorption and its Y-X diagram are illustrated below.

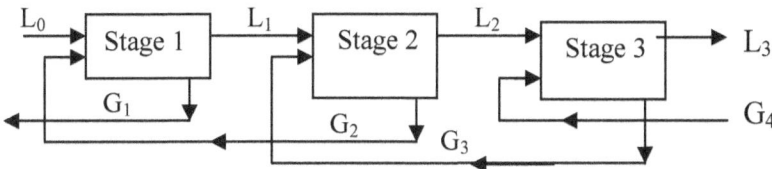

It is usual, though not mandatory, to start counting the number of stages from the point of entrance of the heavier phase. The entering light phase stream, gas stream in the above illustration, would then be G_{n+1} while the heavier phase or liquid stream, entering the 1ST stage would be labelled L_0. If, of course, the gas and liquid streams are defined as free of the diffusing component, then all L_n and all G_n are equal to L and G, respectively.

The corresponding Y-X diagram is illustrated below. The diagram shows three stages, obtained by successively, locating, by means of a horizontal line from Y_{A1}, X_{A0} from the equilibrium line and hence Y_{A2} on the operating line and then repeating the process until the terminal concentrations in the process are reached. The triangle formed in each cycle represents a stage. If there is not an exact match in concentration of this graphical construction with the actual terminal concentrations of the process, the fraction of match of concentration represents the fraction of a stage achieved. In practice, fractional stages are rounded up to a full stage in an intentional over-design.

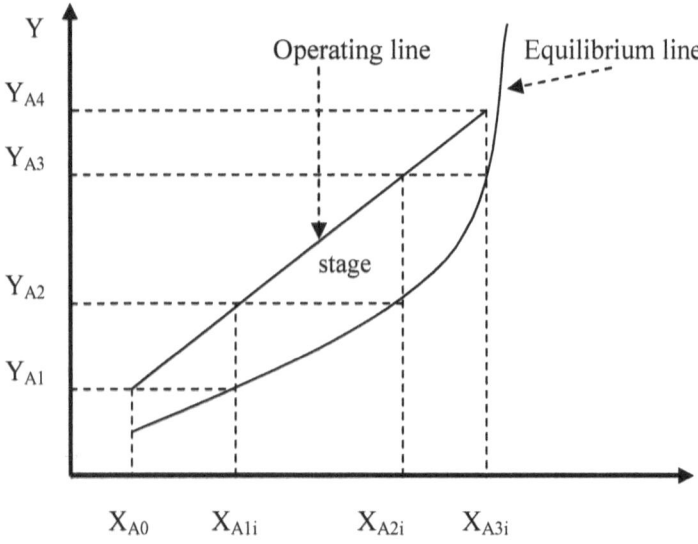

It is, also, usual to define a theoretical stage as that stage in which the diffusing component compositions in its exit streams (the light and heavy phases) are in equilibrium. This concept helps to define the efficiency of a stage in terms of how the separation it achieves compares to that obtained in a theoretical stage.

It should also be noted that not all stagewise operations occur in separate individual units such as in tanks and mixers and as the diagram above may imply. Plate columns are stagewise equipment although all the plates, though separate, are housed in one column. Plate columns are, depicted often, as shown below

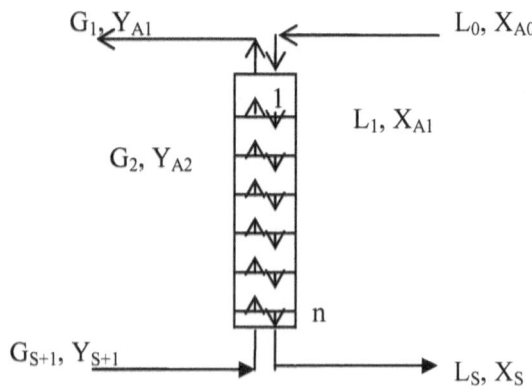

CHAPTER FOUR: MASS TRANSFER WITHOUT CHEMICAL REACTION

Material Balance

The material balance is made on the n^{TH} plate, for the total mass flow and for the total component flow.

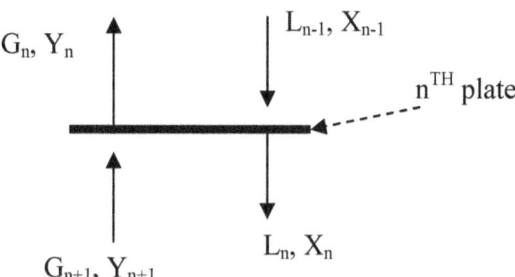

At steady state, no chemical reaction, mole ratios used, L and G independent of concentration of component, A, mass balance on total flow gives a trivial solution

$$G + L = L + G \qquad (4.68)$$

Material balance on component, A, gives

$$Y_n G + X_n L = X_{n-1} L + Y_{n+1} G \qquad (4.69)$$

from which the operating line is obtained as

$$Y_n = \frac{L}{G} X_{n-1} + Y_{n+1} - \frac{L}{G} X_n \qquad (4.70)$$

If the inlet gas and exit liquid concentrations are known, Y_{S+1} and X_S, are known, equation (4.70) becomes easier to work with as

$$Y_n = \frac{L}{G} X_{n-1} + Y_{S+1} - \frac{L}{G} X_S$$

$$= \frac{L}{G} X_{n-1} + b \qquad (4.71)$$

where $b = Y_{S+1} - \frac{L}{G} X_S$ is $cons \tan t$

Estimating the Number of Stages or Plates

This can be done by means of a graphical procedure or by numerical methods. In the graphical procedure, the Y-X diagram is drawn showing the operating and equilibrium lines. Starting from the liquid entry point on the operating curve, such as A, a horizontal line is drawn to intersect the equilibrium curve at B. A vertical line is drawn from this intersection, B, to intersect the operating line again at C and the process is repeated until the exit concentration on the operating line is reached. The number of triangles contiguous with the operating line gives the number of stages. If a fraction of a stage occurs, it is estimated as illustrated in the Baker's method for estimating the number of overall transfer units.

The numerical procedure involves solving the equilibrium equation for X, starting as for the graphical procedure, using this X to solve for a new Y in the operating line equation and repeating the process until terminal concentrations are reached.

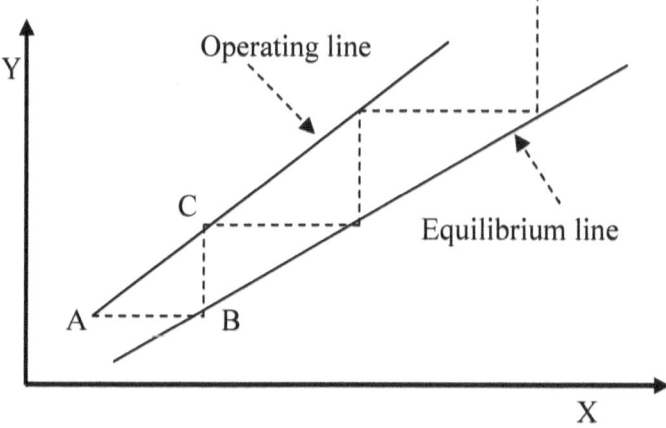

4.3: WORKED EXAMPLES

Example 1

Acetone is to be removed from a gas stream of 10,000 mol/h in a packed column. The inlet gas contains 2.6 mole percent acetone and the outlet gas stream can contain no more than 0.5 mole percent acetone. Assume a pure water stream enters the packed tower at a rate of 8,000 mol/h. What is the concentration of acetone in the outgoing water stream? (*Cheresouirces, 2008*)

Answer(Cheresources, 2008)

The sketch of the process is shown below.

CHAPTER FOUR: MASS TRANSFER WITHOUT CHEMICAL REACTION

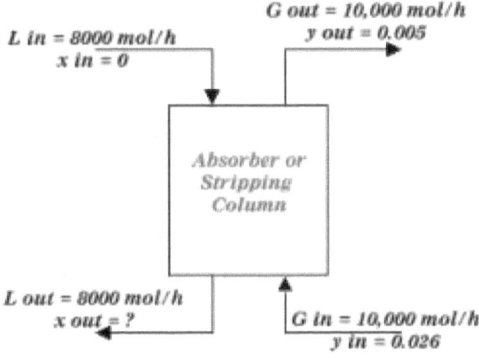

The mass balance is

$$x_{out} \cdot L_{out} + y_{out} \cdot G_{out} = x_{in} \cdot L_{in} + y_{in} \cdot G_{in} \quad (1)$$

That is

$$x_{out} \cdot 8000 + y_{out} \cdot 10000 = 0 \times 8000 + 0.026 \times 10000$$

or $x_{out} = 0.02625$ Ans (2)

Example 2

Specify the packing type and column dimensions for a column that will be used to remove chlorine from a gas stream using an organic solvent. Assume the separation requires 20 theoretical stages. The vapour flow is 7000 kg/h, the average vapour density is 4.8 kg/m^3. The liquid flow is 5000 kg/h, the average liquid density is 833 kg/m^3. The liquid's kinematic viscosity is 0.48 centistokes (4.8 x 10^{-7} m^2/s) (*Cheresources, 2008*)

Answer (*Cheresources, 2008*)

Step 1: Selecting A Type And Size Of Packing

The first consideration is whether to use random or structured packing. Random packing is the type that comes in a sack and is simply dry or wet dumped into the column. Structured packing may come in bales or intricate designs that are stacked in specific patterns. Structured packing is best for very low pressure drop applications and for increasing the capacity of an existing column. Random packing is more economical for new design for which no serious pressure drop is expected. The charts, below, show both English and Metric unit packing factors for the most common random packing types. Under the **Nominal Packing Size, in** heading in column 3, row 3 in the table, are column diameters, also, in inches

Generally, the column diameter to packing size ratio should be greater than 30 for Raschig rings, 15 for ceramic saddles, and 10 for rings or plastic saddles. The geometry of packing is, typically, a function of the required surface area and/or allowable pressure drop. If several packings meet requirements, the least expensive is chosen so long as it has an acceptable operating life.

Packing Factors for Column Packings

Packing Type	Material	Nominal Packing Size, in										
		1/4	3/8	1/2	5/8	3/4	1	1 1/4	1 1/2	2	3	3 1/2
Hy-Pak	Metal						43			18		15
Super Intalox Saddles	Ceramic						60			30		
Super Intalox Saddles	Plastic						33			21		16
Pall Rings	Plastic				97		52		40	24	22	16
Pall Rings	Metal				70		48		33	20	37	16
Intalox Saddles	Ceramic	725	330	200		145	92		52	40		
Raschig Rings	Ceramic	1600	1000	580	380	255	155	125	95	65	32	
Raschig Rings	Metal, 1/32 in	700	390	300	170	155	115					
Raschig Rings	Metal, 1/16 in			410	290	220	137	110	83	57		
Berl Saddles	Ceramic	900		240		170	110		65	45		
Tellerettes	Plastic						38			19		
Mas Pac	Plastic									32		20
Quartz Rock										160		
Cross Partition	Ceramic										80	
Flexipac	Metal						33			22		16
Interlox	Metal						41			27		18
Chempak	Metal						29					

For this example, we'll choose Pall rings (plastic). For columns over 24 inches in diameter, No. 2 or 2 inch packing should be examined first. The value of the applicable flowrate in the problem is often a good guide to the choice of column diameter. For now, we'll settle on 2 inch plastic Pall rings for our initial analysis.

Packing Factors for Column Packings

Packing Type	Material	Nominal Packing Size, mm										
		6.35	9.53	12.70	15.88	19.05	25.40	31.75	38.10	50.80	76.20	88.90
Hy-Pak	Metal						43			18		15
Super Intalox Saddles	Ceramic						60			30		
Super Intalox Saddles	Plastic						33			21		16
Pall Rings	Plastic				97		52		40	24	22	16
Pall Rings	Metal				70		48		33	20	37	16
Intalox Saddles	Ceramic	725	330	200		145	92		52	40		
Raschig Rings	Ceramic	1600	1000	580	380	255	155	125	95	65	32	
Raschig Rings	Metal, .79 mm	700	390	300	170	155	115					
Raschig Rings	Metal, 1.6 mm				410	290	220	137	110	83	57	
Berl Saddles	Ceramic	900		240		170	110		65	45		
Tellerettes	Plastic						38			19		
Mas Pac	Plastic									32		20
Quartz Rock										160		
Cross Partition	Ceramic										80	
Flexipac	Metal						33			22		16
Interlox	Metal						41			27		18
Chempak	Metal						29					

Step 2: Determine the Column Diameter

Most methods for determining the size of randomly packed towers are derived from the Sherwood correlation. A design gas rate, G, can be determined with the help of the figure below which is based on correlation from the Sherwood equation.

Each line on the graph is marked with an acceptable pressure drop in inches of water per foot of packing (numbers in parentheses are in mm of water per meter of packing). Guidelines are as follows:

Moderate to high pressure distillation	= 0.4 to 0.75 in water / ft packing = 32 to 63 mm water / m packing
Vacuum Distillation	= 0.1 to 0.2 in water / ft packing = 8 to 16 mm water / m packing
Absorbers and Strippers	= 0.2 to 0.6 in water / ft packing = 16 to 48 mm water / m packing

These guidelines are designed around "flooding pressure drops" documented in literature. In other words, for most cases, designing with these pressure drops should help you avoid flooding. In the later stages of design, you may want to perform a thorough flooding calculation. Perry's Chemical Engineers'

Handbook covers this topic well. Since we are designing an absorber, we will design for 42 mm water/m packing (you could

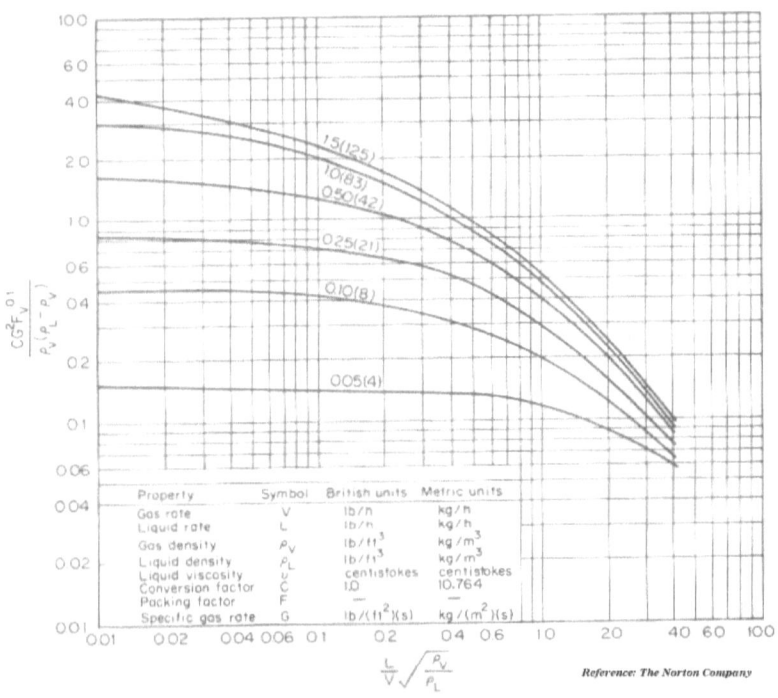

design for a lower pressure drop, but the column will increase in diameter and most likely cost). First, we'll evaluate the x-axis of the graph above as

$$\frac{L}{V} \cdot \left(\frac{Vapour \ Density}{Liquid \ Density} \right)^{0.5} = \frac{5000}{7000} \left(\frac{4.2}{833} \right)^{0.5} = 0.0507 \quad (3)$$

Note that 4.2 kg/m³ was used for the vapour density. The average vapour density was given as 4.8 kg/m³. However, at the top of the column, the vapour will be less dense and at its highest velocity. This is what you should design for. As a rule of thumb, reduce the average vapour density by about 15% for design, in the absence of real data from a similar tower. Reading the intersection of the 42 mm water/m packing line and 0.05 on the axis, we find a value of 1.5 for the y-axis.

CHAPTER FOUR: MASS TRANSFER WITHOUT CHEMICAL REACTION

From the previous charts, we read a packing factor of 24 for 2 inch plastic Pall rings. All other information is known so we can solve for G as shown on the y-axis of the graph:

$$G = \left[1.5 \cdot \frac{(4.2)(833 - 4.2)}{(10.764)(24)(0.48)^{0.1}} \right]^{0.5} = 4.66 \, kg/m^2 \, s \quad (4)$$

Now, we solve for the column cross sectional area:

$$A_x = \frac{Vapour \; Flow}{G}$$

$$= \frac{7000}{(4.66 \, kg/m^2 \, s)(3600 \, s/hr)} = 0.42 \, m^2 \quad (5)$$

and the column diameter is calculated by:

$$Column \; Diameter = \left(\frac{A_x}{\pi/4} \right)^{0.5} = \left(\frac{0.42}{\pi/4} \right)^{0.5} = 0.73 \, m \quad (6)$$

So our assumption of at least a 24 in column diameter is accurate. If it had not been accurate, G would have had to be recalculated using a smaller packing which would also correspond to a larger packing factor.

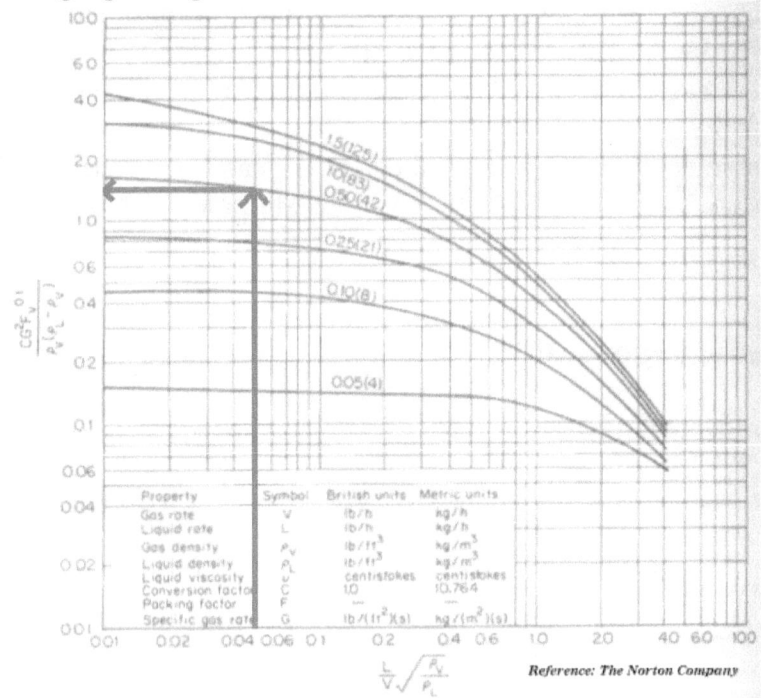

Reference: The Norton Company

Step 3: Determine Column Height

At the preliminary design stage, it is sufficient to determine the general HETP (Height Equivalent to a Theoretical Plate). For more accurate estimation of the height of the column, data from the packing manufacturer must be used. See also the book entitled <u>Distillation Design</u> by Henry Kister (McGraw-Hill, ISBN 0-07-034909-6). <u>Distillation Design</u> contains an exhaustive list of HETP values based on the components of the system and the type of packing used (Chapters 10 and 11). Initial estimate of HETP for absorption/stripping, for all sizes of packing, is about 1.83 m.
To determine the height of the absorption tower in our example, we multiple the 20 theoretical stages by 1.83m to give a preliminary estimate of the height of the tower to be about 37 meters.

Example 3

It is desired to desorb a solute from a solution using a gas. The solution enters a plate column at 20 mole % of solute A and leaves the bottom at 2 mole %. The entering gas is solute free. The slope of the operating line is 1.1 and the equilibrium line is given by $y = x + 0.02$.
Determine the number of theoretical plates and the vapour concentration at each plate. When the Murphree plate efficiency is $E_V = 0.80$, calculate the number of actual plates (*Nguyen, 1979*)

Answer (Nguyen, 2008)

<u>The Operating Line</u>

The material balance around plate, i, in a counter-current plate column is

$$G_1 y_1 + L_i x_i = G_{i-1} y_{i-1} + L_1 x_1 \qquad (7)$$

Assuming constant molal overflow, this may be rearranged as

$$y_{i-1} = \frac{L}{G}x_i + \left(y_1 - \frac{L}{G}x_1\right) \qquad (8)$$

Design Equations

Assuming equilibrium on each plate and a linear relationship between x and y, the equilibrium line may be represented as

$$y_i^* = mx_i + c \qquad (9)$$

where y_i^* is the vapour mole fraction of A in equilibrium with the liquid mole fraction, x, and m is the slope of the equilibrium line. Eliminating x_i in equations (8) and (9)

$$y_{i-1} = \left(\frac{L}{mG}\right)y_i^* - \frac{Lc}{mG} + y_1 - \frac{L}{G}x_1 \qquad (10)$$

The vapour Murphree plate efficiency, E_V, is

$$E_V = \frac{y_i - y_{i-1}}{y_i^* - y_{i-1}} \qquad (11)$$

or

$$y_i^* = \frac{y_i - y_{i-1}}{E_V} + y_{i-1} \qquad (12)$$

Substituting for y_i^* in equation (10) and rearranging, we get that

$$y_i - \left(1 + \frac{mGE_V}{L} - E_V\right)y_{i-1} = E_V\left[mx_1 + c - \left(\frac{mG}{L}\right)y_1\right] \qquad (13)$$

or

$$y_i - C.y_{i-1} = D \qquad (14)$$

where

$$C = \left(1 + \frac{mGE_V}{L} - E_V\right) \quad \text{and}$$

$$D = E_V\left[mx_1 + c - \frac{mG}{L}y_1\right] \qquad (15)$$

Equation (14) is a linear difference equation which can be put in the form

$$y_i = A.C_1^i + \frac{D_2}{1 - C_1} \qquad (16)$$

where A is a constant. Substituting for C and D from equation (15) into equation (16)

$$y_i = A\left[1+E_V\left(\frac{mG}{L}-1\right)\right]^i + \frac{\left(mx_1+c-\frac{mG}{L}y_1\right)}{\left(1-\frac{mG}{L}\right)} \quad (17)$$

When $i = 0$, $y_i = y_1$ A is determined. The complete solution for n actual plates is

$$\left[1+E_V\left(\frac{mG}{L}-1\right)\right]^n = \frac{\left(1-\frac{mG}{L}\right)y_n - mx_1 - c + \frac{mG}{L}y_1}{(y_1 - mx_1 - c)} \quad (18)$$

The material balance for the entire column is (ends labeled 1 and 2)

$$Gy_1 + Lx_2 = Gy_2 + Lx_1 \quad (19)$$

from which

$$x_1 = \frac{G}{L}y_1 - \frac{G}{L}y_2 + x_2 \quad (20)$$

Using equation (20) in equation (18) to eliminate x_1

$$\beta_V^n = \frac{y_1 + \alpha}{y_n + \alpha} \quad (21)$$

where

$$\beta_V = \frac{1}{1+E_V\left(\frac{mG}{L}-1\right)} \quad (22)$$

$$\alpha = \frac{y_2 - \left(\frac{L}{mG}\right)(mx_2+c)}{\left(\frac{L}{mG}\right)-1} \quad (23)$$

Equation (21) is used to calculate the actual composition in each plate. When re-arranged as follows, it is, also, used to calculate the actual number of plates. That is

$$n = \frac{\ln\left(\frac{y_1+\alpha}{y_n+\alpha}\right)}{\ln \beta_V} \quad (24)$$

The number of ideal plates is, since $E_V = 1$,

CHAPTER FOUR: MASS TRANSFER WITHOUT CHEMICAL REACTION

$$n_{ideal} = \frac{\ln\left(\frac{y_1 + \alpha}{y_n + \alpha}\right)}{\ln \lambda} \quad (25)$$

where $\lambda = \dfrac{L}{mG}$. The overall efficiency, E^0, is

$$E^0 = \frac{n_{ideal}}{n} = \frac{\ln \beta_V}{\ln \lambda} = \frac{\ln\left[1 + E_V\left(\dfrac{mG}{L} - 1\right)\right]}{\ln\left(\dfrac{mG}{L}\right)} \quad (26)$$

For the liquid phase, the Murphree plate efficiency is, in terms of the liquid phase,

$$E_L = \frac{x_{i+1} - x_i}{x^*_{i+1} - x_i} \quad (27)$$

Combining equations (27), (7) and (9) with the boundary conditions $i = 0$, $x_i = x_2$ and $i = n$, $x_i = x_n = x_1$

$$x_{i+1} - \left[1 - E_L\left(1 - \frac{L}{mG}\right)\right]x_i = \frac{E_L y_2}{m} - \frac{E_L L x_2}{mG} - \frac{E_L c}{m} \quad (28)$$

That is

$$\beta_L^n = \frac{x_n + \phi}{x_2 + \phi} \quad (29)$$

where

$$\beta_L = 1 + E_L\left(\frac{L}{mG} - 1\right) \quad (30)$$

$$\phi = \frac{x_1 - \left(\dfrac{G}{L}\right)(y_1 - c)}{\left(\dfrac{mG}{L}\right) - 1} \quad (31)$$

and

$$n = \frac{\ln\left(\dfrac{x_n + \phi}{x_2 + \phi}\right)}{\ln \beta_L} \quad (32)$$

$$n_{ideal} = \frac{\ln\left(\frac{x_n + \phi}{x_2 + \phi}\right)}{\ln \lambda} \qquad (33)$$

$$E^0 = \frac{n_{ideal}}{n} = \frac{\ln \beta_L}{\ln \lambda} = \frac{\ln\left[1 + E_L\left(\frac{L}{mG} - 1\right)\right]}{\ln\left(\frac{L}{mG}\right)} \qquad (34)$$

Since the overall efficiency, E^0, of equation (26) is the same as that of equation (34)

$$E^0 = \frac{\ln\left[1 + E_V\left(\frac{mG}{L} - 1\right)\right]}{\ln\left(\frac{mG}{L}\right)} = \frac{\ln\left[1 + E_L\left(\frac{L}{mG} - 1\right)\right]}{\ln\left(\frac{L}{mG}\right)} \qquad (35)$$

from which we get that

$$E_V = \frac{\lambda}{(1-\lambda)[1 + E_L(\lambda - 1)]} - \frac{\lambda}{1-\lambda}$$

$$= \frac{E_L}{E_L + \frac{mG}{L}(1 - E_L)} = \frac{\lambda E_L}{1 + E_L(\lambda - 1)} \qquad (36)$$

and

$$E_L = \frac{E_V}{E_V + \lambda(1 - E_V)} \qquad (37)$$

To determine the <u>number of theoretical plates</u>, with

$y_1 = 0, \quad x_1 = 0.02, \quad x_2 = 0.20, \quad m = 1 \quad \text{and} \quad \frac{L}{G} = 1.1$

From equation (19)

$$y_2 = \frac{L}{G}(x_2 - x_1) + y_1 = 1.1(0.20 - 0.02) + 0 = 0.198 \qquad (38)$$

From equations (22) and (23

$$\beta_V = \frac{1}{1 + 0.80\left(\frac{1}{1.1} - 1\right)} = 1.08 \qquad (39)$$

$$\alpha = \frac{0.198 - 1.1(0.20 + 0.02)}{1.1 - 1} = -0.44 \qquad (40)$$

CHAPTER FOUR: MASS TRANSFER WITHOUT CHEMICAL REACTION

$$\lambda = \frac{L}{mG} = 1.1 \qquad (41)$$

From equations (24), (25) and (26)

$$n = \frac{\ln\left(\dfrac{0-0.44}{0.198-0.44}\right)}{\ln 1.08} = 8.0 \qquad (42)$$

The number of ideal plates is, since $E_V = 1$,

$$n_{ideal} = \frac{\ln\left(\dfrac{0-0.44}{0.198-0.44}\right)}{\ln 1.1} = 6.3 \qquad (43)$$

The overall efficiency, E^0, is

$$E^0 = \frac{n_{ideal}}{n} = \frac{6.3}{8.0} = 0.79 \qquad (44)$$

From equations (24), (25), (39) and (40), we can calculate the theoretical and actual compositions on the plates from

$$y_{n\,ideal} = \frac{y_1 + \alpha}{\lambda^n} - \alpha = \frac{-0.44}{1.1^n} + 0.44 \qquad (45)$$

$$x_{n\,ideal} = y_{n\,ideal} - 0.02 \qquad (46)$$

$$y_n = \frac{y_1 + \alpha}{\beta_V^n} - \alpha = \frac{-0.44}{1.078^n} + 0.44 \qquad (47)$$

$$x_n = y_n - 0.22 \qquad (48)$$

REFERENCES

1. Class Notes at Imperial College of Science & Technology, London, 1965.
2. Cheresources.com; Packed Column Design, 2008
3. Nguyen, Hung Xuan; Calculating Actual Plates in Absorbers and Strippers; Chemical Engineering, April 9, 1979

CHAPTER FIVE
MASS TRANSFER WITH CHEMICAL REACTION

5.0: Introduction

The analysis of mass transfer with chemical reaction becomes complicated because of the fact that there are many varieties of chemical reactions and reaction conditions. The analysis that follows will, therefore, be in terms of well known reaction types especially those that are encountered in industrial practice. These fall, fortunately, into a relatively, few classes or types of reactions namely, very fast uncatalysed reactions, uncatalysed first order irreversible and reversible reactions, some uncatalysed higher order reactions and catalysed reactions.

5.1: Conservation of Mass in Diffusion with Chemical Reaction in Homogenous Medium

Recall that the mass balance over an element, in which mass transfer occurs, was given in Chapter 2 as

$$\frac{\partial C_A}{\partial \theta} = -div.N_A + R_A \qquad (2.11)$$

where R_A is the rate of formation of A per unit volume by chemical reaction and θ is time and, from the same Chapter 2,

$$N_{A_z} = U_z C_A + n_{A_z} + n'_{A_z} \qquad (2.1)$$

$$n_{A_z} = -D_{AB}\frac{\partial C_A}{\partial z} = -D_{AB}C_T\frac{\partial X_A}{\partial z} \qquad (2.2)$$

$$n'_{A_z} = -\varepsilon_D\frac{\partial C_A}{\partial z} = -\varepsilon_D C_T\frac{\partial X_A}{\partial z} \qquad (2.3)$$

Combining equations (2.2) and (2.3) with equation (2.1)

$$N_{A_z} = U_z C_A - (D_{AB} + \varepsilon_D)\frac{\partial C_A}{\partial z}$$

$$= U_z C_T X_A - (D_{AB} + \varepsilon_D)C_T\frac{\partial X_A}{\partial z} \qquad (2.4)$$

where

C_T = total concentration, kmol/m^3

C_A = molar concentration of A, kmol/m³
$U_z C_A$ = convective flux, kmol/m².s
n_{AZ} = molar flux of A by molecular diffusion in the z direction, kmol/m².s
n^t_{AZ} = molar flux of A by eddy diffusion in the z direction, kmol/m².s
D_{AB} = molecular diffusivity of A through the medium, B, m²/s
X_A = mole fraction of A
ε_D = eddy diffusivity of A through B, m²/s

When eddy diffusion is not significant and chemical reaction controls the mass transfer, $\varepsilon_D = 0$ and, since $C_A = X_A C_T$, equation (2.4) becomes

$$N_{A_z} = U_z C_A - D_{AB} \frac{\partial C_A}{\partial z} \qquad (5.1)$$

Substituting equation (5.1) into equation (2.11) and simplifying, we got equation (2.12) below such that

$$\frac{\partial C_A}{\partial \theta} + U_z \frac{\partial C_A}{\partial z} = D_{AB} \frac{\partial^2 C_A}{\partial z^2} - C_A \frac{\partial U_z}{\partial z} + R_A \qquad (2.12)$$

where, in three dimensions,

$$div.P = \left[\frac{\partial P_x}{\partial x} + \frac{\partial P_y}{\partial y} + \frac{\partial P_z}{\partial z}\right] \quad P \text{ can be any function} \quad (2.13)$$

For a stationary medium, $U_z = 0$ and $\frac{\partial U_z}{\partial z} = 0$ and

$$\frac{\partial C_A}{\partial \theta} = D_{AB}\left[\frac{\partial^2 C_A}{\partial z^2}\right] + R_A \qquad (5.2)$$

It can be seen, because of the presence of the time element, θ, that equation (5.1) can be used in the penetration theory analysis of mass transfer with chemical reaction. For steady state mass transfer, $\frac{\partial C_A}{\partial \theta} = 0$ and equation (5.2) reduces to

$$D_{AB}\left[\frac{\partial^2 C_A}{\partial z^2}\right] + R_A = 0 \qquad (5.3)$$

Equation (5.3) would be the basic equation for the Whitman film theory analysis of mass transfer with chemical reaction.

CHAPTER FIVE: MASS TRANSFER WITH CHEMICAL REACTION

5.2: Chemical Kinetics

If we consider two reactants A and B which react in solvent S to form products P and Q, the chemical reaction equation can be expressed as

$$aA + bB \leftrightarrow pP + qQ \qquad (5.4)$$

and a, b, p and q are the stoichiometric coefficients. The rate of reaction with respect to A, R_A, is defined as the rate of formation of A per unit volume of solution and is generally, a function of temperature, pressure and the reactants in the solution. Mathematically,

$$R_A = f(T) \cdot f(C_A, C_B, C_P, C_Q) \qquad (5.5)$$

For an irreversible reaction

$$aA + bB \rightarrow pP + qQ \qquad (5.6)$$

$$R_A = -k_j C_A^m C_B^n \qquad (5.7)$$

where

$$j = m + n = \text{order of the reaction} \qquad (5.8)$$

Note that m and n may or may not be equal to the stoichiometric coefficients, a and b. Table 5.1, below, illustrates the various forms of the reaction equation for various, commonly, encountered orders of reaction.

Table 5.1: Forms of Rate Equations for Orders of Reaction

Order of Reaction	Form of Rate Equation	
Zero Order	$R_A = -k_0$	(5.9)
First Order	$R_A = -k_1 C_A$	(5.10)
Second Order	$R_A = -k_2 C_A C_B$ or $R_A = -k_2 C_A^2$ etc	(5.11)
Third Order	$R_A = -k_2 C_A^3$ or $R_A = -k_2 C_A C_B^2$ etc	(5.12)
n^{TH} Order	$R_A = -k_n C_A^n$	(5.12a)

Note, also, that

$$R_B = \frac{b}{a} R_A \qquad (5.13)$$

5.3: The General Case of Diffusion with Chemical Reaction on the Basis of the Whitman Film Theory.

When equation (5.7) is substituted into equation (5.2) or (5.3), the general equations to be solved, for either the unsteady state or steady state mass transfer, are obtained. Thus, for the reaction stated in equation (5.4) having the rate equation (5.7), equations (5.2) and (5.3) become, respectively,

$$\frac{\partial C_A}{\partial \theta} = D_{AS} \frac{\partial^2 C_A}{\partial z^2} - k_j C_A^m C_B^n \qquad (5.14)$$

$$\frac{\partial C_B}{\partial \theta} = D_{BS} \frac{\partial^2 C_B}{\partial z^2} - \frac{b}{a} k_j C_A C_B \qquad (5.15)$$

for unsteady state (penetration theory) conditions, and

$$D_{AS} \frac{\partial^2 C_A}{\partial z^2} - k_j C_A^m C_B^n = 0 \qquad (5.16)$$

$$D_{BS} \frac{\partial^2 C_B}{\partial z^2} - \frac{b}{a} k_j C_A^m C_B^n = 0 \qquad (5.17)$$

for steady state (Whitman film theory) conditions. These equations are solved simultaneously. It is important, also, to remember that, basically,

$$N_{AG} = N_{AL} \qquad (5.18)$$

that is, the molar flux in the gas phase is always equal to the molar flux in the liquid phase. And that, for transfer from the gas to the liquid phase (gas absorption)

$$N_{AG} = k_G (p_{A_b} - p_{A_i}) \qquad (5.19)$$

and

$$p_{Ai} = H C_{Ai} \qquad (5.20)$$

where k_G and p_{Ab} are constant, H is Henry's law constant and $p_{Ai} < p_{Ab}$.

It is, also, helpful to compare, visually, what happens at the interface when there is no chemical reaction and when there is chemical reaction. The broken lines in the gas and liquid phases, in Figure 5.1, show the pressure and concentration profiles in physical absorption (that is, no chemical reaction) and in absorption with chemical reaction.

Figure 5.1: Partial Pressure and Concentration Profiles in the Vicinity of the Interface for Gas Absorption with and without Chemical Reaction

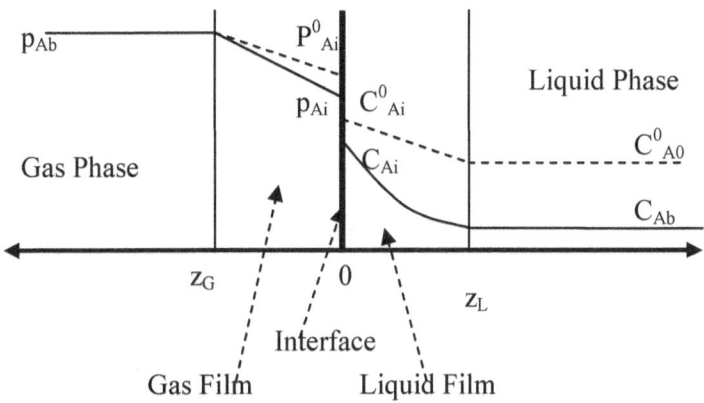

Figure 5.2: Partial Pressure and Concentration Profiles in the Vicinity of the Interface for Gas Absorption with Very Fast Chemical Reaction

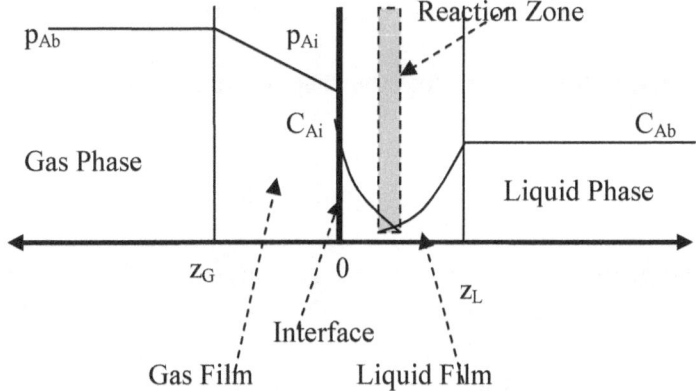

For slow reactions, the reaction zone is large while for very fast reactions, the reaction zone becomes a plane and mass transfer is due to diffusion through this plane with the condition that

$$bN_A + aN_B = 0 \quad i.e. \quad \frac{N_A}{N_B} = -\frac{a}{b} \qquad (5.21)$$

Figure 5.3: Partial Pressure and Concentration Profiles in the Vicinity of the Interface for Gas Absorption with Slow Chemical Reaction

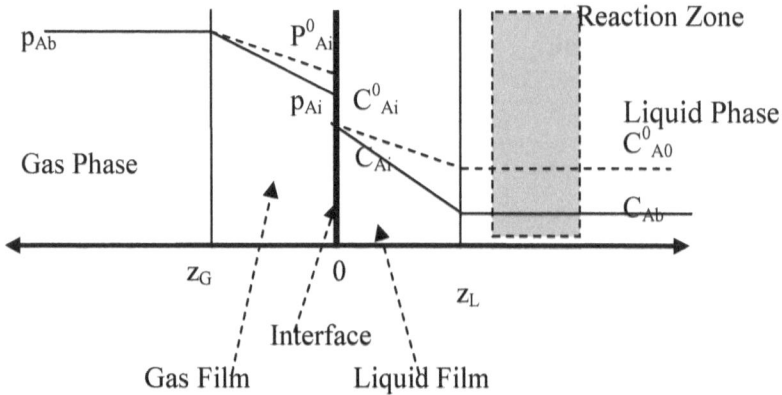

It is seen that the reaction zone is in the liquid film while, for slow reactions, it is assumed that reaction occurs in the bulk of the liquid. This implies that diffusion and reaction occur in series in slow reactions.

5.3.1: Special Cases of Diffusion with Chemical Reaction on the Basis of the Whitman Film Theory: Steady State Gas Absorption With Very Fast Chemical Reaction

Figure 5.4, below, shows that the reaction plane moves, in a finite amount of time, from the interface at $z = 0$ and $\theta = 0$ to $z = z_R$ and $\theta = \theta$. It, also, shows that at $z = z_R$, $C_{AR} = 0$. . It is assumed that the interface is at equilibrium. That is

$$p_{Ai} = HC_{Ai} \qquad (5.20)$$

Assuming, also, that the solution is dilute, that D_A and D_B are independent of each other and that the reaction, in solvent, S, is

$$aA + bB \leftrightarrow pP + qQ \qquad (5.4)$$

we can analyse the situation using some of the results we obtained for physical absorption

CHAPTER FIVE: MASS TRANSFER WITH CHEMICAL REACTION

Figure 5.4: Partial Pressure and Concentration Profiles in the Vicinity of the Interface for Gas Absorption with Very Fast Chemical Reaction

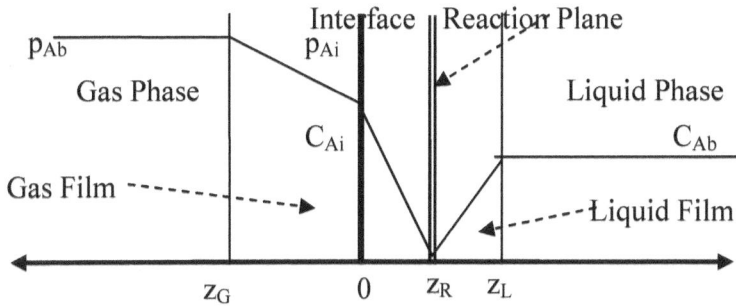

Recall that, in physical absorption, the liquid film and the gas phase based overall mass transfer coefficients were found to be

$$k_L = \frac{D_{AS}}{z_L} \quad (5.22)$$

$$\frac{1}{K_{OG}} = \frac{1}{k_G} + \frac{H}{k_L} \quad (5.23)$$

where k_G, k_L are the gas and liquid film mass transfer coefficients, respectively, K_{OG} the overall mass transfer coefficient based on the gas phase, and H the Henry's law constant, equivalent to m, the slope of the equilibrium line. We found, also, that

$$N_A = k_L(C_{A_i} - C_{A_b}) \quad \text{where} \quad k_L = \frac{D_{AB}}{L} \quad (2.19)$$

If we replace C_{Ab} by $C_{AR} = 0$ and substitute equation (5.22) in equation (2.19), we get that

$$N_A = \frac{D_{AS}}{z_R} C_{A_i} = k_L \frac{z_L}{z_R} C_{A_i} \quad (5.24)$$

For diffusion in the liquid side of the reaction zone

$$N_B = -\frac{D_{BS}}{(z_L - z_R)} C_{B_{bi}} = -k_L \frac{D_{BS}}{D_{AS}} \frac{C_{B_b}}{1 - \frac{z_R}{z_L}} \quad (5.25)$$

From stoichiometry, equation (5.21),

$$b N_A = -a N_B \quad (5.26)$$

From equations (5.26), (5.25) and (5.24), since $N_A = -N_B$, we get that

$$b\frac{z_L}{z_R}C_{A_i} = a\frac{D_{BS}}{D_{AS}}\frac{C_{B_b}}{\left(1 - \frac{z_R}{z_L}\right)} \quad (5.27)$$

This gives, on rearrangement,

$$\frac{z_L}{z_R} - 1 = \frac{a}{b}\frac{D_{BS}}{D_{AS}}\frac{C_{B_b}}{C_{A_i}} \quad (5.28)$$

or

$$C_{A_i}\frac{z_L}{z_R} = C_{A_i} + \frac{a}{b}\frac{D_{BS}}{D_{AS}}C_{B_b} \quad (5.29)$$

Substituting equation (5.29) into equation (5.24)

$$N_A = k_L\left(C_{A_i} + \frac{a}{b}\frac{D_{BS}}{D_{AS}}C_{B_b}\right) \quad (5.30)$$

This is the equation for molar flux for very fast chemical reaction. When this equation is compared to equation (2.19) which describes the molar flux for physical absorption only, it can be seen that while the driving force for physical absorption decreases, by $-C_{Ab}$, that for chemical reaction increases by $\frac{a}{b}\frac{D_{BS}}{D_{AS}}C_{B_b}$.

Similar relationships can be developed for gas phase based mass transfer for very fast chemical reactions.

For physical absorption in the gas phase, equation (2.19) becomes

$$N_A = k_G(p_{Ab} - p_{Ai}) \quad (5.31)$$

From equations (5.31) and (5.20)

$$\frac{N_A}{k_G} = p_{Ab} - HC_{Ai} = p_{A_b} - H\left(\frac{N_A}{k_L} - \frac{a}{b}\frac{D_{BS}}{D_{AS}}C_{B_b}\right) \quad (5.32)$$

That is

$$N_A\left(\frac{1}{k_G} + \frac{H}{k_L}\right) = p_{A_b} + H\frac{a}{b}\frac{D_{BS}}{D_{AS}}C_{B_b} \quad (5.33)$$

or

$$N_A = K_{OG}\left(p_{A_b} + H\frac{a}{b}\frac{D_{BS}}{D_{AS}}C_{B_b}\right) \quad (5.34)$$

The comparable expression for physical absorption, based on the

CHAPTER FIVE: MASS TRANSFER WITH CHEMICAL REACTION

gas phase, is

$$N_A = K_{OG}(p_{Ab} - HC_{Ai}) \tag{5.35}$$

Critical Value of C_{Bb}

It is helpful to determine the maximum, theoretical value of B, $C_{Bb}*$, in the liquid phase for any given constant partial pressure of A in the gas phase for which

$$N_A = K_{OG}\left(p_{A_b} + H\frac{a}{b}\frac{D_{BS}}{D_{AS}}C_{B_b}\right) \tag{5.34}$$

if $C_{Bb} < C_{Bb}*$ and

$$N_A = k_G p_{A_b} \tag{5.36}$$

if $C_{Bb} \geq C_{Bb}*$

This approach is illustrated, schematically, as shown below

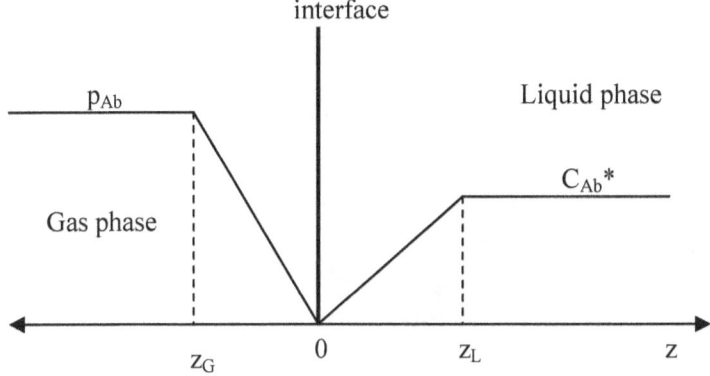

From equation (5.25)

$$N_B = -k_L \frac{D_{BS}}{D_{AS}} C_{Bb}^* \tag{5.37}$$

From equation (5.36)

$$N_A = k_G p_{Ab} \tag{5.38}$$

$C_{Bb}*$ is obtained from equations (5.26), (5.25) and (5.36) as

$$b k_G p_{Ab} = +a k_L \frac{D_{BS}}{D_{AS}} C_{B_b} \tag{5.39}$$

so that

$$C^*_{Bb} = \frac{b\,k_G\,D_{AS}}{a\,k_L\,D_{BS}} p_{A_b} \quad (5.40)$$

Similarly, p_{Ab}^* can be obtained as

$$p^*_{Ab} = \frac{a\,k_L\,D_{BS}}{b\,k_G\,D_{AS}} C_{B_b} \quad (5.41)$$

Theoretical versus Actual Predictions

The figure below illustrates the comparison between theoretical prediction and actual observation.

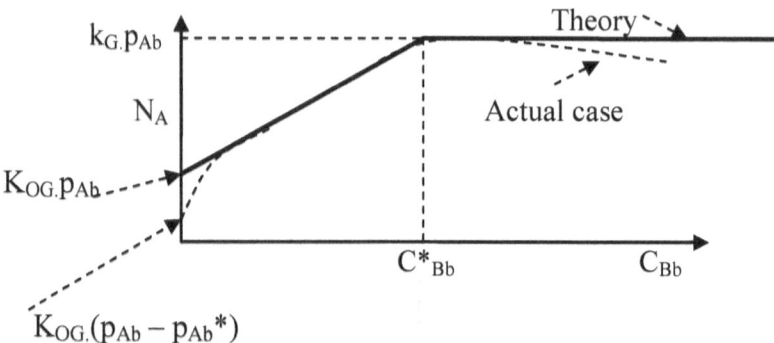

The drop observed in the actual case is due to changes in liquid properties such as viscosity, etc, as a result of absorption and reaction.

5.3.1.1: Calculation of the Height of a Packed Tower

Consider a packed tower with a section shown, schematically, below. Let

G_m = Molar flow rate of gas phase (constant for dilute mixture)
L_V = Volumetric flow rate of liquid phase
π = Total operating pressure
S = Interfacial area
A = Cross-sectional area of column

By a material balance over the section shown

$$-G_m \frac{dp_A}{\pi} = N_A\,dS = \frac{a}{b} L_V\,dC_B \quad (5.42)$$

CHAPTER FIVE: MASS TRANSFER WITH CHEMICAL REACTION

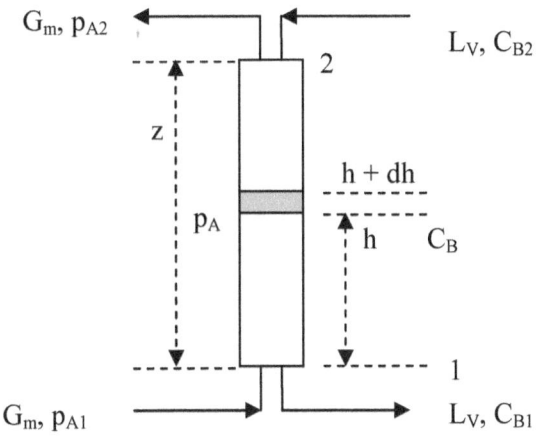

But
$$dS = a_S \, A \, dh \quad (5.43)$$
where a_S is the interfacial area per unit volume of packing. Substituting equation (5.43) into (5.42) and re-arranging, we get that
$$z = \int_0^z dh = \frac{G_m}{\pi a_S A} \int_{p_{A2}}^{p_{A1}} \frac{d\,p_A}{N_A} \quad (5.44)$$
By a material balance between entry point 1 and differential section shown
$$\frac{G_m}{\pi}(p_{A1} - p_A) = \frac{a}{b} L_V (C_B - C_{B1}) \quad (5.45)$$
from which we get that
$$p_A = -\frac{a}{b}\frac{L_V \pi}{G_m} C_B + p_{A1} + \frac{a}{b}\frac{L_V \pi}{G_m} C_{B1} \quad (5.46)$$
the so called stoichiometric operating line. Thus, if

<u>$C_{B1} \geq C_{B1}^*$; $C_{B2} > C_{B2}^*$: Case I</u>
$$N_A = k_G \, p_{Ab} \quad (5.38) \text{ and}$$
$$z = \int_0^z dh = \frac{G_m}{\pi a_S A}\int_{p_{A2}}^{p_{A1}} \frac{d\,p_A}{N_A} = H_G \ln \frac{p_{A_1}}{p_{A_2}} \quad (5.47)$$

<u>$C_{B1} < C_{B1}^*$; $C_{B2} < C_{B2}^*$: Case II</u>
$$N_A = K_{OG}\left(p_{A_b} + H \frac{a}{b}\frac{D_{BS}}{D_{AS}} C_{B_b}\right) \quad (5.34)$$

and

$$z = \int_0^{\xi} dh = \frac{G_m}{\pi a_S A K_{OG}} \int_{p_{A2}}^{p_{A1}} \frac{dp_A}{p_{A_b} + H\frac{a}{b}\frac{D_{BS}}{D_{AS}}C_{B_b}} \qquad (5.48)$$

From the stoichiometric operating line, equation (5.46)

$$C_B = -\frac{b}{a}\frac{G_m}{\pi L_V}p_A + C_{B1} + \frac{b}{a}\frac{G_m}{\pi L_V}p_{A1} \qquad (5.49)$$

so that

$$p_A + H\frac{a}{b}\frac{D_{BS}}{D_{AS}}C_{B_b} = p_A - \frac{1}{\alpha}\frac{D_{BS}}{D_{AS}}p_A + \frac{1}{\alpha}\frac{D_{BS}}{D_{AS}}p_{A_1}$$

$$+ H\frac{a}{b}\frac{D_{BS}}{D_{AS}}C_{B_1} \qquad (5.50)$$

$$p_A + H\frac{a}{b}\frac{D_{BS}}{D_{AS}}C_{B_b} = \left[1 - \frac{1}{\alpha}\frac{D_{BS}}{D_{AS}}\right]p_A$$

$$+ \left[\frac{1}{\alpha}\frac{D_{BS}}{D_{AS}}p_{A_1} + H\frac{a}{b}\frac{D_{BS}}{D_{AS}}C_{B_1}\right] \qquad (5.50a)$$

where α = absorption factor = $\frac{\pi L_V}{H G_m}$ (5.50b). Equation (5.50a) can be re-stated as

$$p_A + H\frac{a}{b}\frac{D_{BS}}{D_{AS}}C_{B_b} = m p_A + q$$

$$\text{where } m = \left[1 - \frac{1}{\alpha}\frac{D_{BS}}{D_{AS}}\right] \text{ and }$$

$$q = \left[\frac{1}{\alpha}\frac{D_{BS}}{D_{AS}}p_{A_1} + H\frac{a}{b}\frac{D_{BS}}{D_{AS}}C_{B_1}\right] \qquad (5.51)$$

Substituting equation (5.51) into equation (5.48), integrating and re-arranging, we get that

$$z = \frac{H_{OG}}{1 - \frac{1}{\alpha}\frac{D_{BS}}{D_{AS}}} \ln\frac{\left(p_A + H\frac{a}{b}\frac{D_{BS}}{D_{AS}}C_B\right)_1}{\left(p_A + H\frac{a}{b}\frac{D_{BS}}{D_{AS}}C_B\right)_2} \qquad (5.52)$$

CHAPTER FIVE: MASS TRANSFER WITH CHEMICAL REACTION

$\underline{C_{B1} < C_{B1}^* \; ; \; C_{B2} > C_{B2}^*}$: Case III

Here, from equations (5.47) and (5.53),

$$z = H_G \ln \frac{p_{A_X}}{p_{A_2}} + \frac{H_{OG}}{1 - \frac{1}{\alpha}\frac{D_{BS}}{D_{AS}}} \ln \frac{\left(p_A + H \frac{a}{b} \frac{D_{BS}}{D_{AS}} C_B\right)_1}{\left(p_A + H \frac{a}{b} \frac{D_{BS}}{D_{AS}} C_B\right)_X} \quad (5.53)$$

where X is the point of intersection of the stoichiometric operating line and the line of the critical liquid phase concentration of reactant B. These cases are illustrated, diagramatically, below.

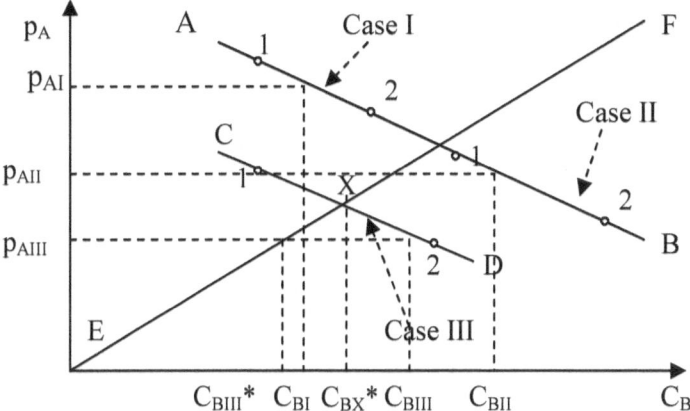

Line EF is the critical partial pressure, p_A^* versus critical concentration, C_B^* line. Lines AB and CD represent general stoichiometric operating lines on which are located specific operating lines between points 1 and 2 in a tower for cases I, II and III. It is seen that case I lies above the critical line, case II below it while case III straddles it. The stoichiometric operating line for case III crosses the critical concentration line at X.
Recall, from equations (5.39) and (5.49), that

$$\frac{b}{a} \frac{k_G}{k_L} \frac{D_{AS}}{D_{BS}} p_{A_X} = -\frac{b}{a} \frac{G_m}{\pi L_V} p_{A_X} + C_{B1} + \frac{b}{a} \frac{G_m}{\pi L_V} p_{A1} \quad (5.54)$$

from which we get that

$$P_{A_X} = \frac{P_{A_1} + \dfrac{a\,\pi L_V}{b\,G_m}C_{B_1}}{1 + \dfrac{\pi L_V}{G_m}\dfrac{D_{AS}}{D_{BS}}\dfrac{k_G}{k_L}} \qquad (5.55)$$

and

$$C_{B_X} = \frac{b}{a}\left[\frac{P_{A_1} + \dfrac{a\,\pi L_V}{b\,G_m}C_{B_1}}{\dfrac{\pi L_V}{G_m} + \dfrac{D_{BS}}{D_{AS}}\dfrac{k_L}{k_G}}\right] \qquad (5.56)$$

5.3.2: Steady State Gas Absorption With Slow Chemical Reaction

Consider the schematic of the tower as shown below

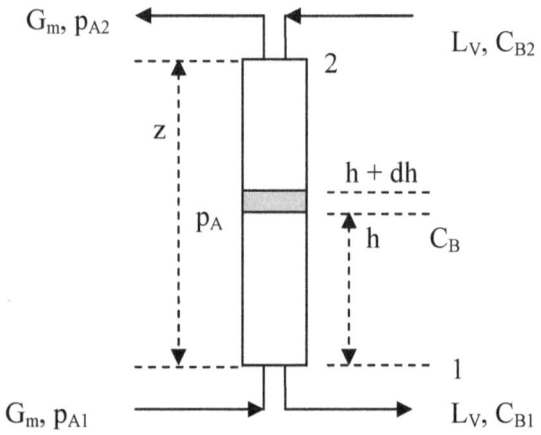

For a slow, irreversible, chemical reaction, the material balance is

$$-G_m \frac{d\,p_A}{\pi} = N_A\,a_S\,A\,dh = -R_A\,H_O\,A\,dh - L_V\,dC_A$$

$$= \frac{a}{b}L_V dC_B - L_V dC_A \qquad (5.57)$$

where H_O is the specific hold-up per unit volume of packed bed, R_A, the rate of reaction per unit volume of solution and N_A is that for physical absorption, equal to $K_{OG}\,(p_A - HC_A)$. Thus, from equation (5.57)

CHAPTER FIVE: MASS TRANSFER WITH CHEMICAL REACTION

$$z = \int_0^z dh = \frac{G_m}{\pi a_S A} \int_{p_{A2}}^{p_{A1}} \frac{dp_A}{N_A} = H_{OG} \int \frac{dp_A}{p_A - HC_A} \quad (5.58)$$

Integration of equation (5.58) is not straightforward since the relationship between p_A and C_A depends on h. To determine this relationship, consider, for example, a slow, irreversible, first order reaction. Its rate of reaction would be given as

$$R_A = -k_1 C_A \quad (5.59)$$

At the top of the column, where relatively fresh and solute free liquid is entering, C_A would be small, k_1 is small and hence R_A would be negligible. This situation would correspond, approximately, to physical absorption for which, from equation (5.57)

$$G_m \frac{dp_A}{\pi} = L_V dC_A \quad (5.60)$$

At the bottom of the tower, concentrations have increased in the liquid stream but, because the reaction is slow, the amount absorbed is approximately, equal to the amount reacted. That is, $L_V dC_A$ is negligible. We call this state of affairs, the pseudo-stationary state. Thus, from equation (5.57)

$$N_A a_S = -R_A H_O \quad (5.61)$$

Equations (5.60) and (5.61) imply that the top and bottom of the column behave differently and may be analysed accordingly. Thus, we can divide the column into two regions where

1. Top of column $G_m \dfrac{dp_A}{\pi} = L_V dC_A$ Gives z_P
2. Bottom of column $K_{OG}(p_A - HC_A)a_S = k_1 C_A H_O$ Gives z_R

z_P is the height of the top region of the tower where physical absorption dominates and z_R is the height of the bottom region of the tower in which chemical reaction dominates. The total height of the absorption tower, z, is given by

$$z = z_P + z_R \quad (5.62)$$

Estimating z_P in the Physical Absorption Zone

Since $G_m \dfrac{dp_A}{\pi} = L_V dC_A$, the material balance expands to

$$\frac{G_m}{\pi}(p_A - P_{A_2}) = L_V(C_A - C_{A_2}) \qquad (5.63)$$

over a small differential element between plane X to plane 2, to which physical absorption applies. For $C_{A2} = 0$, we get that

$$C_A = \frac{G_m}{\pi L_V}(p_A - p_{A_2}) \qquad (5.64)$$

Substituting equation (5.64) into equation (5.58)

$$z_P = H_{OG} \int_{P_{AX}}^{P_{A_2}} \frac{dp_A}{p_A\left(1 - \frac{HG_m}{\pi L_V}\right) + \frac{HG_m}{\pi L_V}p_{A_2}} \qquad (5.65)$$

From equation (5.50b), the absorption factor, α, was defined as

$$\alpha = \frac{\pi L_V}{H G_m}. \text{ Hence}$$

$$z_P = H_{OG}\int_{P_{AX}}^{P_{A_2}} \frac{dp_A}{p_A\left(1 - \frac{1}{\alpha}\right) + \frac{p_{A_2}}{\alpha}}$$

$$= \frac{H_{OG}}{1 - \frac{1}{\alpha}}\ln\left[\frac{\left(1 - \frac{1}{\alpha}\right)p_{AX} + \frac{p_{A_2}}{\alpha}}{\left(1 - \frac{1}{\alpha}\right)p_{A_2} + \frac{p_{A_2}}{\alpha}}\right]$$

$$= \frac{H_{OG}}{1 - \frac{1}{\alpha}}\ln\left[\left(1 - \frac{1}{\alpha}\right)\frac{p_{AX}}{p_{A_2}} + \frac{1}{\alpha}\right] \qquad (5.66)$$

Estimating z_R in the Pseudo-stationary Absorption Zone

From equation (5.61)

$$N_A a_S = -R_A H_O \qquad (5.61)$$

or

$$K_{OG}(p_A - HC_A)a_S = -R_A H_O \qquad (5.67)$$

That is

$$p_A - HC_A = -\frac{H_O}{K_{OG} a_S}R_A \qquad (5.68)$$

Substituting equation (5.68) into equation (5.58) we get that

CHAPTER FIVE: MASS TRANSFER WITH CHEMICAL REACTION

$$z_R = \frac{H_{OG} K_{OG} a_S}{H_O} \int_{p_{A_1}}^{p_{AX}} \frac{dp_A}{R_A} \qquad (5.69)$$

5.3.2.1: Zero Order Reactions

For a zero order reaction,
$$R_A = -k_0 \qquad (5.70)$$
and
$$z_R = -\frac{H_{OG} K_{OG} a_S}{H_O} \int_{p_{A_1}}^{p_{AX}} \frac{dp_A}{k_0} \qquad (5.71)$$

But from equation (5.68), p_A is related to C_A as
$$p_A - HC_A = -\frac{H_O k_0}{K_{OG} a_S} = \beta_0 \qquad (5.72)$$

Thus
$$z_R = -\frac{H_{OG} K_{OG} a_S}{H_O} \int_{p_{A_1}}^{p_{AX}} \frac{dp_A}{k_0}$$
$$= \frac{H_{OG}}{\beta_0}(p_{A_1} - p_{AX}) \qquad (5.73)$$

At the plane, X, on which physical absorption ends and chemical reaction begins, equation (5.72) becomes
$$p_{AX} - HC_{AX} = -\frac{H_O k_0}{K_{OG} a_S} = \beta_0 \qquad (5.74)$$

while the material balance becomes
$$\frac{G_m}{\pi}(p_{AX} - P_{A_2}) = L_V(C_{AX} - C_{A_2}) \qquad (5.75)$$

from which, since C_{A2} was assumed to be zero,
$$HC_{AX} = \frac{HG_m}{\pi L_V}(p_{AX} - P_{A_2}) = \frac{1}{\alpha}(p_{AX} - P_{A_2}) \qquad (5.76)$$

Substituting equation (5.76) into equation (5.74)
$$p_{AX}\left(1 - \frac{1}{\alpha}\right) = \beta_0 - \frac{P_{A_2}}{\alpha} \qquad (5.77)$$

Equation (5.77) gives the value of p_{AX} for the evaluation of equations (5.66) and (5.73). Thus, putting equation (5.77) into equation (5.66), we get

$$z_P = \frac{H_{OG}}{\left(1 - \dfrac{1}{\alpha}\right)} \ln\left[\frac{\beta_0}{p_{A_2}}\right] \tag{5.78}$$

and into equation (5.73) we get

$$z_R = \frac{H_{OG}}{\beta_0}\left[p_{A_1} - \frac{\left(\beta_0 - \dfrac{p_{A_2}}{\alpha}\right)}{\left(1 - \dfrac{1}{\alpha}\right)}\right] \tag{5.79}$$

and
$$z = z_P + z_R \tag{5.62}$$

5.3.2.2: First Order Reactions

For a first order reaction,
$$R_A = -k_1 C_A \tag{5.80}$$
and
$$z_R = -\frac{H_{OG} K_{OG} a_S}{k_1 H_O} \int_{p_{A_1}}^{p_{AX}} \frac{dp_A}{C_A} \tag{5.81}$$

At the plane, X, on which physical absorption ends and chemical reaction begins, equation (5.68) becomes

$$\frac{p_{AX}}{H} - C_{AX} = \frac{H_O k_1 C_{AX}}{H K_{OG} a_S} = \beta_1 C_{AX}$$

or
$$C_{AX} = \frac{p_{AX}}{(1 + \beta_1)} \tag{5.82}$$

where
$$\beta_1 = \frac{H_O k_1}{H K_{OG} a_S} \tag{5.83}$$

Substituting equation (5.82) into equation (5.81)

$$z_R = \frac{H_{OG}(1 + \beta_1)}{\beta_1} \ln \frac{p_{A_1}}{p_{AX}} \tag{5.84}$$

From equations (5.75) and (5.82)

$$p_{AX} - p_{A_2} = \frac{\pi L_V}{G_m} C_{AX} = \frac{\pi L_V}{G_m} \frac{p_{AX}}{(1 + \beta_1)} = \frac{\alpha}{1 + \beta_1} p_{AX}$$

CHAPTER FIVE: MASS TRANSFER WITH CHEMICAL REACTION

or
$$p_{A_X} = \frac{p_{A_2}}{1 - \dfrac{\alpha}{1 + \beta_1}} \qquad (5.85)$$

Putting equation (5.85) into equation (5.84)

$$z_R = \frac{H_{OG}(1 + \beta_1)}{\beta_1} \ln\left[\left(1 - \frac{\alpha}{1 + \beta_1}\right)\frac{p_{A_1}}{p_{A_2}}\right] \qquad (5.86)$$

and into equation (5.66)

$$z_P = \frac{H_{OG}}{1 - \dfrac{1}{\alpha}} \ln\left[\left(1 - \frac{1}{\alpha}\right)\frac{1}{1 - \dfrac{\alpha}{1 + \beta_1}} + \frac{1}{\alpha}\right] \qquad (5.87)$$

and
$$z = z_P + z_R \qquad (5.62)$$

5.3.2.3: Second Order Reactions

For a second order reaction,
$$R_A = -k_2 C_A^2 \qquad (5.88)$$

and
$$z_R = -\frac{H_{OG} K_{OG} a_S}{k_2 H_O} \int_{p_{A_1}}^{p_{A_X}} \frac{d p_A}{C_A^2} \qquad (5.89)$$

From equations (5.68) and (5.88)

$$p_A - HC_A = -\frac{H_O}{K_{OG} a_S} k_2 C_A^2 = \beta_2 (HC_A)^2$$

or
$$p_A = HC_A + \beta_2 (HC_A)^2 \qquad (5.90)$$

where
$$\beta_2 = \frac{H_O k_2}{H^2 K_{OG} a_S} \qquad (5.91)$$

Equation (5.89) becomes

$$z_R = \frac{H_{OG}}{\beta_2} \int_{p_{AX}}^{p_{A_1}} \frac{d p_A}{(HC_A)^2} \qquad (5.92)$$

From equation (5.90)
$$d p_A = d(HC_A) + 2\beta_2 (HC_A) d(HC_A)$$
$$= [1 + 2\beta_2 (HC_A)] d(HC_A) \qquad (5.93)$$

Substituting equation (5.93) into equation (5.92) and integrating

$$z_R = \frac{H_{OG}}{\beta_2} \int_{p_{AX}}^{p_{A_1}} \frac{[1 + 2\beta_2 (HC_A)] d(HC_A)}{(HC_A)^2}$$

$$= \frac{H_{OG}}{\beta_2} \int_{p_{AX}}^{p_{A_1}} \frac{d(HC_A)}{(HC_A)^2} + \frac{H_{OG}}{\beta_2} \int_{p_{AX}}^{p_{A_1}} \frac{2\beta_2 d(HC_A)}{(HC_A)}$$

$$= H_{OG} \left[\frac{1}{H\beta_2} \left(\frac{1}{C_{AX}} - \frac{1}{C_{A_1}} \right) + 2\ln \frac{C_{A_1}}{C_{AX}} \right] \quad (5.94)$$

At the plane, X, on which physical absorption ends and chemical reaction begins, equation (5.90) in equation (5.76) becomes

$$\beta_2 (HC_{A_X})^2 + (1 - \alpha)(HC_{A_X}) - P_{A_2} = 0 \quad (5.95)$$

which gives

$$HC_{A_X} = \frac{(\alpha - 1) \pm \sqrt{(1-\alpha)^2 + 4\beta_2 P_{A_2}}}{2\beta_2} \quad (5.96)$$

and again, from equation (5.76) and (5.96)

$$HC_{A_X} = \frac{(\alpha - 1) \pm \sqrt{(1-\alpha)^2 + 4\beta_2 P_{A_2}}}{2\beta_2} = \frac{1}{\alpha}(p_{AX} - P_{A_2})$$

or $$p_{AX} = P_{A_2} + \alpha \frac{(\alpha - 1) \pm \sqrt{(1-\alpha)^2 + 4\beta_2 P_{A_2}}}{2\beta_2} \quad (5.97)$$

Substituting for p_{AX} in equation (5.66) we get that

$$z_P = \frac{H_{OG}}{1 - \frac{1}{\alpha}} \ln \left[\left(1 - \frac{1}{\alpha}\right) \left(1 + \alpha \frac{(\alpha - 1) \pm \sqrt{(1-\alpha)^2 + 4\beta_2 P_{A_2}}}{2\beta_2 P_{A_2}} \right) + \frac{1}{\alpha} \right] \quad (5.98)$$

and $$z = z_P + z_R \quad (5.62)$$

5.3.3: Steady State Gas Absorption With Comparable Rates of Molecular Diffusion and Chemical Reaction

The equation to be solved is

$$D_{AB} \left[\frac{\partial^2 C_A}{\partial z^2} \right] + R_A = 0 \quad (5.3)$$

with the boundary conditions

CHAPTER FIVE: MASS TRANSFER WITH CHEMICAL REACTION

$z = 0 \qquad\qquad C_A = C_{Ai}$
$z = z_L \qquad\qquad C_A = C_{Ab}$

The molar flux is, still, from equations (5.19) and (5.20),
$$N_{AG} = k_G(p_{A_b} - p_{A_i}) \qquad (5.19)$$
and
$$p_{Ai} = HC_{Ai} \qquad (5.20)$$

5.3.3.1: Zero Order Reactions

For a zero order reaction,
$$R_A = -k_0 \qquad (5.70)$$
and equation (5.3) becomes
$$\frac{d^2C_A}{dz^2} - \frac{k_0}{D_{AS}} = 0 \qquad (5.99)$$

Integrating twice
$$\frac{dC_A}{dz} = \frac{k_0}{D_{AS}}z + A \qquad (5.100)$$

$$C_A = \frac{k_0}{2D_{AS}}z^2 + Az + B \qquad (5.101)$$

and putting in the boundary conditions, we get that

$$C_A = C_{A_i} - \left[\frac{k_L}{D_{AS}}(C_{A_i} - C_{A_b}) + \frac{k_0}{2k_L}\right]z + \frac{k_0}{2D_{AS}}z^2 \qquad (5.102)$$

$$N_{Ai} = -D_{AS}\left(\frac{dC_A}{dz}\right)_{z=0} = -D_{AS}\left[-\frac{k_L}{D_{AS}}(C_{A_i} - C_{A_b}) - \frac{k_0}{2k_L}\right]$$

$$= k_L\left[(C_{A_i} - C_{A_b}) + \frac{k_0 D_{AS}}{2k_L^2}\right] \qquad (5.103)$$

Since $\qquad \dfrac{1}{K_{OG}} = \dfrac{1}{k_G} + \dfrac{H}{k_L} \qquad (5.23)$

$$N_{Ai} = K_{OG}\left[p_{A_b} - H\left(C_{A_b} - \frac{k_0 D_{AS}}{2k_L^2}\right)\right] \tag{5.104}$$

Note that the reaction is slow if $C_{A_b} \gg \frac{k_0 D_{AS}}{2k_L^2}$.

The material balance is

$$-G_m \frac{dp_A}{\pi} = K_{OG}\left[p_{A_b} - H\left(C_{A_b} - \frac{k_0 D_{AS}}{2k_L^2}\right)\right] a_S A \, dh$$

$$= k_0 H_O A \, dh - L_V \, dC_A \tag{5.105}$$

The height of packing is calculated by stepwise integration of equation (5.105).

5.3.3.2: First Order Reactions

For a first order reaction,

$$R_A = -k_1 C_A \tag{5.106}$$

and equation (5.106) becomes

$$D_{AS} \frac{d^2 C_A}{dz^2} - k_1 C_A = 0 \tag{5.107}$$

or

$$\frac{d^2 C_A}{dz^2} - q^2 C_A = 0 \tag{5.108}$$

where

$$q = \sqrt{\frac{k_1}{D_{AS}}} \tag{5.109}$$

Integrating equation (5.109)

$$C_A = A' \sinh(qz) + B' \cosh(-qz) \tag{5.110}$$

and putting in the boundary conditions, we get that

$$C_A = \frac{C_A \sinh[q(z_L - z)] + C_{A_b} \sinh(qz)}{\sinh q z_L} \tag{5.111}$$

$$N_{Ai} = -D_{AS}\left(\frac{dC_A}{dz}\right)_{z=0}$$

$$= -\frac{D_{AS}}{\sinh(qz_L)}\left[C_{Ai}\cosh(qz_L)x(-q) + C_{Ab}qx\cosh(qx0)\right]$$

$$= \frac{qD_{AS}}{\sinh(qz_L)}\left[C_{Ai}\cosh(qz_L) - C_{Ab}\right] \quad (5.112)$$

Equation (5.112) may be further rearranged to show that

$$N_{Ai} = \frac{qD_{AS}}{\sinh(qz_L)}\left[C_{Ai}\cosh(qz_L) - C_{Ab}\right]$$

$$= \frac{z_L}{z_L}qD_{AS}\coth(qz_L)\left[C_{Ai} - \frac{C_{Ab}}{\cosh(qz_L)}\right]$$

$$= k_L\frac{qz_L}{\tanh(qz_L)}\left[C_{Ai} - \frac{C_{Ab}}{\cosh(qz_L)}\right] \quad (5.113)$$

If we define $qz_L = \beta$, then

$$N_{Ai} = \frac{k_L\beta}{\tanh(\beta)}\left[C_{Ai} - \frac{C_{Ab}}{\cosh(\beta)}\right] \quad (5.114)$$

$\dfrac{\beta}{\tanh(\beta)}$ is known as the Hatta number, Ha or N_{Ha}. The significance of β may be illustrated as follows

$$\beta = qz_L = \sqrt{\frac{k_1}{D_{AS}}}\cdot z_L = \sqrt{\frac{k_1 z_L^2}{D_{AS}}}$$

$$= \frac{\sqrt{k_1 D_{AS}}}{K_L}$$

$$= \frac{M.T.C \text{ for fast 1st order reaction}}{M.T.C. \text{ for physical absorption}} \quad (5.115)$$

From equations (5.19), (5.20) and (5.114) and since

$$\frac{1}{K_{OG}} = \frac{1}{k_G} + \frac{H}{k_L} \quad (5.23)$$

$$N_{Ai} = Ha.k_L\left[C_{Ai} - \frac{C_{Ab}}{\cosh(\beta)}\right] = k_G(p_{Ab} - p_{Ai})$$

$$N_{Ai} = (K_{OG})_R\left[p_{Ab} - \frac{H}{\cosh\beta}C_{Ab}\right] \quad (5.116)$$

where

$$\frac{1}{(K_{OG})_R} = \frac{1}{k_G} + \frac{H}{k_L \cdot Ha} \qquad (5.117)$$

The material balance is

$$-G_m \frac{dp_A}{\pi} = (K_{OG})_R \left(p_{A_b} - \frac{H}{\cosh \beta} \right) a_S A \, dh$$

$$= k_1 C_A H_O A \, dh - L_V \, dC_A \qquad (5.118)$$

The height of packing is calculated by stepwise integration of equation (5.118).

Special Cases: Slow Reaction (β is small)
Since

$$\cosh \beta = \frac{e^\beta + e^{-\beta}}{2} = \frac{1 + \beta + \frac{\beta^2}{2} + 1 - \beta + \frac{\beta^2}{2}}{2}$$

$$= 1 + \frac{\beta^2}{2} \qquad (5.119)$$

$$\tanh \beta = \frac{e^\beta - e^{-\beta}}{e^\beta + e^{-\beta}} = \frac{\left(1 + \beta + \frac{\beta^2}{2}\right) - \left(1 - \beta + \frac{\beta^2}{2}\right)}{\left(1 + \beta + \frac{\beta^2}{2}\right) + \left(1 - \beta + \frac{\beta^2}{2}\right)}$$

$$= \frac{\beta}{1 + \frac{\beta}{2}} \qquad (5.120)$$

reaction is slow if $\beta^2 \ll 1$ and the height of packing is given by

$$z_R = \frac{H_{OG}(1 + \beta_1)}{\beta_1} \ln \left[\left(1 - \frac{\alpha}{1 + \beta_1}\right) \frac{p_{A_1}}{p_{A_2}} \right] \qquad (5.86)$$

and into equation (5.66)

$$z_P = \frac{H_{OG}}{1 - \frac{1}{\alpha}} \ln \left[\left(1 - \frac{1}{\alpha}\right) \frac{1}{1 - \frac{\alpha}{1 + \beta_1}} + \frac{1}{\alpha} \right] \qquad (5.87)$$

and
$$z = z_P + z_R \qquad (5.62)$$

CHAPTER FIVE: MASS TRANSFER WITH CHEMICAL REACTION

Note that β and β_1 are not the same.

Special Cases: Fast Reaction (β is large)

$$\cosh \beta = \frac{e^\beta + e^{-\beta}}{2} = \frac{e^\beta}{2}$$

$$\frac{1}{\cosh \beta} = 2e^{-\beta} \to 0 \qquad (5.121)$$

$$\tanh \beta = \frac{e^\beta - e^{-\beta}}{e^\beta + e^{-\beta}} = 1$$

$$Ha = \beta = \frac{\sqrt{k_1 D_{AS}}}{k_L} \qquad (5.122)$$

Reaction is fast if $e^{-\beta} \ll e^\beta$ and $e^\beta \gg 2$. For example, when $\beta = 3$, $e^3 = 20.08$, $e^{-3} = 0.05$ so that for $\beta \geq 3$,

$$N_{AG} = (K_{OG})_R P_{A_b} \qquad (5.123)$$

where

$$\frac{1}{(K_{OG})_R} = \frac{1}{k_G} + \frac{H}{\sqrt{k_L D_{AS}}} \qquad (5.117)$$

The height of the packed tower is obtained from the material balance

$$-G_m \frac{dp_A}{\pi} = (K_{OG})_R p_A a_S A d h \qquad (5.124)$$

as

$$z = (H_{OG})_R \ln \frac{p_{A_1}}{p_{A_2}} \qquad (5.125)$$

where

$$(H_{OG})_R = H_G + \frac{H G_m}{L_V}(H_L)_R \qquad (5.126)$$

$$(H_L)_R = \frac{L_V}{(k_L)_R a_S A} = \frac{L_V}{a_S A \sqrt{k_1 D_{AS}}} \qquad (5.127)$$

A plot of the molar flux versus β shows the regions where pure physical absorption with gas film or liquid film resistance controlling the mass transfer and where chemical reaction controls.

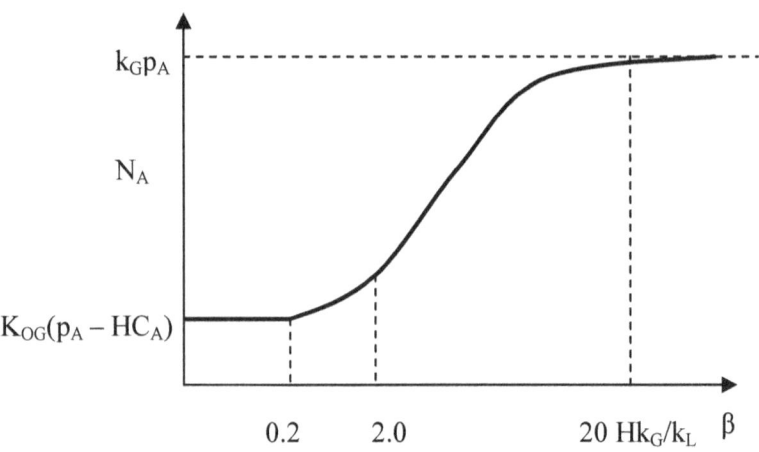

5.3.3.3: The General Case of Diffusion with an n^{TH} Order Chemical Reaction on the Basis of the Whitman Film Theory.

For an n^{TH} order reaction,

$$R_A = -k_n C_A^n \tag{5.128}$$

and equation (5.128) becomes

$$D_{AS} \frac{d^2 C_A}{dz^2} - k_n C_A^n = 0 \tag{5.129}$$

or

$$\frac{d^2 C_A}{dz^2} - \frac{k_n}{D_{AS}} C_A^n = 0 \tag{5.130}$$

Multiplying equation (5.130) throughout by $2 \dfrac{dC_A}{dz}$

$$2 \frac{dC_A}{dz} \frac{d^2 C_A}{dz^2} - \frac{k_n}{D_{AS}} 2 \frac{dC_A}{dz} C_A^n = 0 \quad i.e. \quad \frac{d}{dz}\left(\frac{dC_A}{dz}\right)^2 = \frac{2k_n}{D_{AS}} C_A^n \frac{dC_A}{dz}$$

$$or \quad d\left(\frac{dC_A}{dz}\right)^2 = \frac{2k_n}{D_{AS}} C_A^n dC_A \tag{5.131}$$

and integrating, we get

CHAPTER FIVE: MASS TRANSFER WITH CHEMICAL REACTION

$$\left(\frac{dC_A}{dz}\right)^2 = \frac{2k_n}{D_{AS}}\left(\frac{C_A^{n+1}}{n+1} + A\right) \tag{5.132}$$

where A is a constant of integration. Thus

$$\frac{dC_A}{dz} = \sqrt{\frac{2k_n}{D_{AS}}\left(\frac{C_A^{n+1}}{n+1} + A\right)^{1/2}} \tag{5.133}$$

or

$$\int \frac{dC_A}{\left(\frac{C_A^{n+1}}{n+1} + A\right)^{1/2}} = \sqrt{\frac{2k_n}{D_{AS}}}\, z + B \tag{5.134}$$

and B is another constant. Analytical or numerical integration of equation (5.134) gives the concentration profile and hence N_A using the boundary conditions

$$z = 0 \qquad C_A = C_{Ai}$$
$$z = z_L \qquad C_A = C_{Ab}$$

5.3.3.4: Molecular Diffusion with a Second Order Chemical Reaction with Rate Equation, $R_A = -k_2 C_A C_B$ on the Basis of the Whitman Film Theory.

The applicable equation is

$$D_{AB}\left[\frac{\partial^2 C_A}{\partial z^2}\right] + R_A = 0 \tag{5.3}$$

which for the two reactants, A and B, becomes

$$D_{AS}\frac{d^2 C_A}{dz^2} + R_A = 0 \tag{5.135}$$

$$D_{BS}\frac{d^2 C_B}{dz^2} + R_B = 0 \tag{5.136}$$

From stoichiometry,

$$R_B = \frac{b}{a} R_A \tag{5.13}$$

so that, from equations (5.135), (5.136) and (5.13),

$$D_{AS}\frac{d^2 C_A}{dz^2} = \frac{a}{b} D_{BS} \frac{d^2 C_B}{dz^2} = -R_A$$

$$= k_2 C_A C_B \qquad (5.137)$$

The boundary conditions are

i $z = 0$ $C_A = C_{Ai}$
ii $z = z_L$ $C_A = C_{Ab}$
iii $z = 0$ $\dfrac{dC_B}{dz} = 0 \qquad (5.138)$

iv $z = z_L$ $C_B = C_{Bb}$

It is, also, assumed that reactant B is not volatile and is not transferred across the interface.

The solution of equation (5.137) is made much easier if we define the following dimensionless variables and parameters.

$$\overline{C_A} = \frac{C_A}{C_{Ai}} \qquad (5.139)$$

$$\overline{C_B} = \frac{a}{b}\frac{D_{BS}}{D_{AS}}\frac{C_B}{C_{Ai}} \qquad (5.140)$$

$$\overline{z} = \frac{z}{z_L} \qquad (5.141)$$

Equation (5.137) becomes

$$\frac{d^2 \overline{C_A}}{d\overline{z}^2} = \frac{d^2 \overline{C_B}}{d\overline{z}^2} = \frac{k_2}{D_{AS}^2}\overline{C_A} C_{Ai} \frac{b}{a} \frac{D_{AS}^2}{D_{BS}} \overline{C_B} C_{Ai} \frac{z_L^2}{C_{Ai}}$$

$$= \frac{b\, k_2}{a\, k_L^2} \frac{D_{AS}^2}{D_{BS}} C_{Ai}. \overline{C_A}.\overline{C_B} = \psi \overline{C_A}.\overline{C_B} \qquad (5.142)$$

where

$$\psi = \frac{b\, k_2}{a\, k_L^2}\frac{D_{AS}^2}{D_{BS}} C_{Ai} \qquad (5.143)$$

The dimensionless form of the boundary conditions becomes

CHAPTER FIVE: MASS TRANSFER WITH CHEMICAL REACTION

i $\quad \bar{z} = 0 \quad\quad \overline{C_A} = 1$

ii $\quad \bar{z} = 1 \quad\quad \overline{C_A} = \dfrac{C_{Ab}}{C_{A_i}}$

iii $\quad \bar{z} = 0 \quad\quad \dfrac{d\overline{C_B}}{d\bar{z}} = 0 \quad\quad\quad (5.144)$

iv $\quad \bar{z} = 1 \quad\quad \overline{C_B} = \dfrac{a\,D_{BS}\,C_{B_b}}{b\,D_{AS}\,C_{A_i}}$

An illustrative sketch of the relationships between $\overline{C_A}$, $\overline{C_B}$ and $\bar{z}$ is shown below.

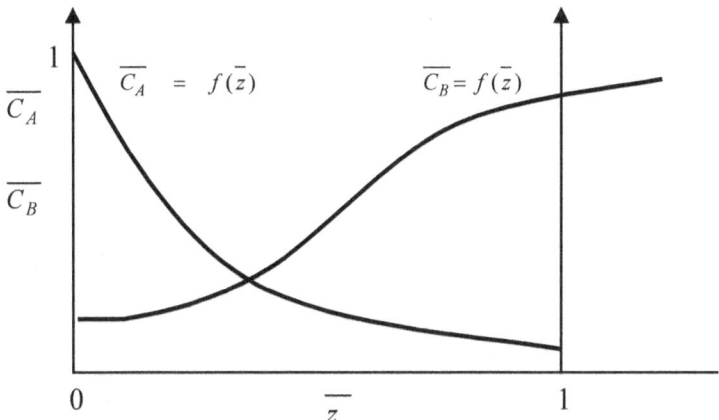

Equation (4.143) is best solved numerically. The relevant equations, for the numerical solution, are

$$\overline{C_A}_{j+1} = \overline{C_A}_j + \left.\dfrac{d\overline{C_A}}{d\bar{z}}\right|_j \Delta\bar{z}_j \quad\quad (5.145)$$

$$\overline{C_B}_{j+1} = \overline{C_B}_j + \left.\dfrac{d\overline{C_B}}{d\bar{z}}\right|_j \Delta\bar{z}_j \quad\quad (5.146)$$

Since

$$\left(\dfrac{d^2\overline{C_A}}{d\bar{z}^2}\right)_j = \left(\dfrac{d^2\overline{C_B}}{d\bar{z}^2}\right)_j = \psi\,\overline{C_A}_j.\overline{C_B}_j \quad\quad (5.142a)$$

175

$$\left(\frac{d\overline{C_A}}{d\overline{z}}\right)_{j+1} = \left(\frac{d\overline{C_A}}{d\overline{z}}\right)_j + \left(\frac{d^2\overline{C_A}}{d\overline{z}^2}\right)_j \Delta\overline{z}_j$$

$$= \left(\frac{d\overline{C_A}}{d\overline{z}}\right)_j + \psi\overline{C_{Aj}}\cdot\overline{C_{Bj}} \quad (5.147)$$

$$\left(\frac{d\overline{C_B}}{d\overline{z}}\right)_{j+1} = \left(\frac{d\overline{C_B}}{d\overline{z}}\right)_j + \left(\frac{d^2\overline{C_B}}{d\overline{z}^2}\right)_j \Delta\overline{z}_j$$

$$= \left(\frac{d\overline{C_B}}{d\overline{z}}\right)_j + \psi\overline{C_{Aj}}\cdot\overline{C_{Bj}} \quad (5.148)$$

<u>Calculation Procedure</u>

i   Start at $\overline{z} = 0$

ii   This gives $(\overline{C_A})_0 = 1$ and $\left.\dfrac{d\overline{C_B}}{d\overline{z}}\right|_0 = 0$

iii Assume $\left.\dfrac{d\overline{C_A}}{d\overline{z}}\right|_0$ and $(\overline{C_B})_0$

iv
$$N_{Ai} = -\left.\frac{dC_A}{dz}\right|_0 = -\frac{D_{AS}}{z_L}C_{Ai}\left.\frac{d\overline{C_A}}{d\overline{z}}\right|_0$$

$$= -k_L C_{Ai}\left.\frac{d\overline{C_A}}{d\overline{z}}\right|_0 = k_G(p_{A_b} - HC_{A_I}) \quad (5.149)$$

v   Calculate C_{Ai} from equation (5.149) and hence ψ from equation (5.143)

vi Calculate $\left(\dfrac{d^2\overline{C_A}}{d\overline{z}^2}\right)_0 = \left(\dfrac{d^2\overline{C_B}}{d\overline{z}^2}\right)_0$ from equation (5.142a)

vii Select $\Delta\overline{z}_0$ and calculate $\left.\dfrac{d\overline{C_A}}{d\overline{z}}\right|_1$ and $\left.\dfrac{d\overline{C_B}}{d\overline{z}}\right|_1$ from equations (5.147) and (5.148)

viii Calculate $\overline{C_{A1}}$ and $\overline{C_{B1}}$ from equations (5.145) and (5.146)

CHAPTER FIVE: MASS TRANSFER WITH CHEMICAL REACTION

ix   Calculate $\left(\dfrac{d^2 \overline{C_A}}{d\overline{z}^2}\right)_1 = \left(\dfrac{d^2 \overline{C_B}}{d\overline{z}^2}\right)_1$ from equation (5.142a)

x   Continue until $(\overline{z})_{j+1} = 1$. If $\overline{C_A} = \dfrac{C_{Ab}}{C_{A_i}}$ and

$\overline{C_B} = \dfrac{a}{b}\dfrac{D_{BS}}{D_{AS}}\dfrac{C_{B_b}}{C_{A_i}}$, stop. Otherwise, repeat iteration with new

assumed values of $\left.\dfrac{d\overline{C_A}}{d\overline{z}}\right|_0$ and $(\overline{C_B})_0$.

Estimation of Column Height; Rigorous Solution

The material balance is

$$-G_m \dfrac{dp_A}{\pi} = N_{A_i} a_S A dh = -R_A H_O A dh - L_V dC_A$$

$$= \dfrac{a}{b} L_V dC_B - L_V dC_A \qquad (5.57)$$

Equation (5.57) can be expressed in finite difference form as

$$-G_m \dfrac{(p_{Aj+1} - p_{Aj})}{\pi} = (N_{A_i})_{averasge}\, a_S A \Delta h_j$$

$$= -k_2 (C_A C_B)_{average}\, H_O A \Delta h_j$$

$$\qquad - L_V (C_{Aj+1} - C_{Aj})$$

$$= \dfrac{a}{b} L_V (C_{Bj} - C_{Bj+1})$$

$$\qquad + L_V (C_{Aj+1} - C_{Aj}) \quad (5.150)$$

Equation (5.150) is solved numerically. Start at the top of the column.

i p_{Aj-1}, C_{Aj-1}, C_{Bj-1} are known. Choose p_{Aj}
ii   First approximation

$$(N_A)_{average} = (N_A)_{j-1} = (p_A - HC_A)_{j-1}$$
$$(C_A C_B)_{average} = C_{Aj-1} C_{Bj-1} \qquad (5.151)$$

iii
$$\Delta h_j = \frac{H_{OG}(p_{Aj} - p_{Aj-1})}{(p_A - HC_A)_{j-1}}$$

$$C_{Aj} = C_{Aj-1} + \frac{G_m}{\pi L_V}(p_{Aj} - p_{Aj-1}) - k_2(C_A C_B)_{j-1} H_O A \frac{\Delta h_j}{L_V}$$

$$C_{Bj} = C_{Bj-1} - \frac{b}{a}\frac{G_m}{\pi L_V}(p_{Aj} - p_{Aj-1}) + \frac{b}{a}(C_{Aj} - C_{Aj-1}) \quad (5.152)$$

iv Second approximation

$$(p_A - HC_A)_{average} = \frac{1}{2}\left[(C_A C_B)_j - (C_A C_B)_{j-1}\right]$$

$$(C_A C_B)_{average} = \frac{1}{2}\left[(p_A - HC_A)_j - (p_A - HC_A)_{j-1}\right]$$

v Continue iteration until there is no significant change in Δh_j, C_{Aj} and C_{Bj}.

vi Proceed to the next step by calculating p_{Aj+1} and hence C_{Aj+1} and C_{Bj+1} from equations (5.150) and (5.151) and repeat the iterations

vii Calculate $N_{Ai\,j+1} = k_G(p_{Aj+1} - HC_{Aj+1})$ until terminal conditions are reached.

Estimation of Column Height; Approximate Solution

An approximate solution, useful in many instances, can be obtained by assuming that the reaction is pseudo first order. That is, instead of the rate equation being $R_A = -k_2 C_A C_B$, it is now represented as $R_A = -k`_1 C_A$ where $k`_1 = k_2 C_B$.
From equation (5.114) and (5.115)

$$N_{Ai} = \frac{k_L \beta}{\tanh(\beta)}\left[C_{Ai} - \frac{C_{Ab}}{\cosh(\beta)}\right]$$

$$= \frac{D_{AS}}{z_R}(C_{Ai} - C_{Ab})$$

and $\beta = \dfrac{\sqrt{k`_1 D_{AS}}}{k_L} = \dfrac{\sqrt{k_1 C_{Bi} D_{AS}}}{k_L} \qquad (5.153)$

CHAPTER FIVE: MASS TRANSFER WITH CHEMICAL REACTION

C_{Bi} is unknown. We can simplify equation (5.153) to

$$\frac{z_L}{z_R}(C_{Ai} - C_{Ab}) = Ha\left[C_{Ai} - \frac{C_{Ab}}{\cosh(\beta)}\right]$$

$$\text{where} \quad Ha = \frac{\beta}{\tanh(\beta)} \quad (5.154)$$

and obtain C_{Bi} by combining equations (5.21) and (5.25) as follows

$$N_A + \frac{a}{b}N_B = 0$$

$$= \frac{D_{AS}}{z_R}(C_{Ai} - C_{Ab})$$

$$+ \frac{a}{b}\frac{D_{BS}}{(z_L - z_R)}(C_{Ai} - C_{Bb}) \quad (5.155)$$

Finally, we can evaluate N_{Ai} as

$$N_{Ai} = k_L \frac{z_L}{z_R}(C_{Ai} - C_{Ab}) \quad (5.156)$$

5.3.4: Estimating Mass Transfer with Chemical Reaction in Plate Columns

A typical representation of a typical element in a tray is illustrated below

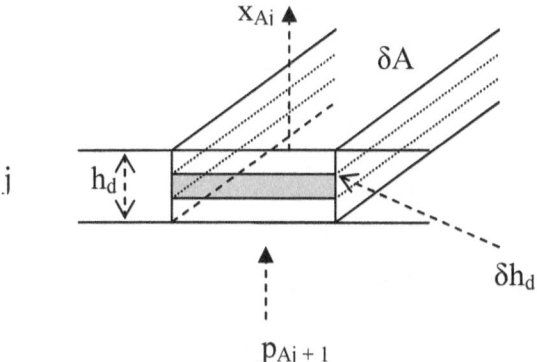

It is usual to assume a dispersion model of fluid behaviour (perfectly mixed fluid) and that point values of variables represent the values of those variables on the tray.

Assuming plug flow of the gas stream, a material balance on component, A, gives

$$-G_m \frac{\delta A}{A} \frac{d p_A}{\pi} = N_A a_D \delta A d h$$

$$= K_{OG}(p_A - HC_A) a_D A d h \quad (5.157)$$

from which we get that

$$-\int_{p_{Aj+1}}^{p_{Aj}} \frac{d p_A}{p_A - HC_A} = \int_0^{h_D} \frac{\pi K_{OG} a_D A}{G_m} d h \quad (5.158)$$

Since liquid is well mixed on the tray, C_A is constant, so that

$$\ln \frac{p_{Aj+1} - HC_A}{p_{Aj} - HC_A} = \frac{\pi K_{OG} a_D A h_D}{G_m} \quad (5.159)$$

Note that reaction did not feature in this analysis. This is because of the choice of δA which we can make as small as possible thus obviating the need to consider chemical reaction.

If we, now, consider the whole tray in relation to other trays in the column, we have to take chemical reaction into account and use a larger, rather than an element, representation of a tray as shown below.

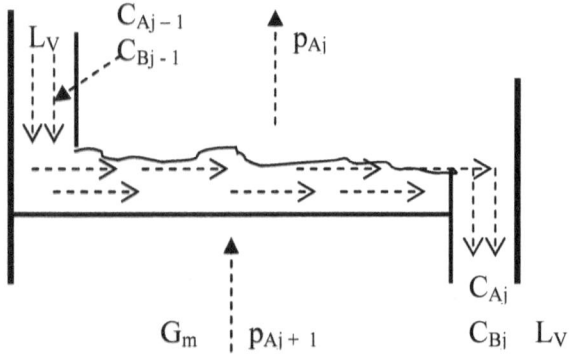

The overall material balance is

$$\frac{G_m (p_{Aj+1} - p_{Aj})}{\pi} = k_2 (C_A C_B)_{average} \varphi_L A d h$$

$$+ L_V (C_{Aj} - C_{Aj-1}) \quad (5.160)$$

where

φ_L = fraction of liquid in dispersion

$$\varphi_L = \frac{volume\ of\ liquid}{volume\ of\ dispersion} \qquad (5.161)$$

In terms of the component, B, the mass balance is, also,

$$k_2 (C_A C_B)_{average} \varphi_L A d h = \frac{a}{b} L_V (C_{Bj-1} - C_{Bj}) \qquad (5.162)$$

It is, usually, assumed that both the liquid and gas are well mixed on the tray, so that p_{Aj+1}, C_{Aj} and C_{Bj} are, all, constant on each tray. Starting from the top of the column, with p_{Aj}, C_{Aj-1} and C_{Bj-1}, equations (5.159), (5.160) and (5.162) can be solved for p_{Aj+1}, C_{Aj} and C_{Bj}.

5.4: Unsteady State Diffusion with Chemical Reaction on the Basis of the Penetration Theory.

5.4.1: Gas Absorption with Very Fast Reaction

Figures 5.1, 5.2 and 5.3 apply. We, still, assume dilute solutions and conditions of zero net flux, that is, there is no convective flux in the direction of mass transfer. If we consider the reaction, taking place in solvent, S,

$$aA + bB = pP \qquad (5.163)$$

The equation to be solved is

$$\frac{\partial C_A}{\partial \theta} = -div.N_A + R_A \qquad (2.11)$$

where R_A is the rate of formation of A per unit volume by chemical reaction and θ is time. For a dilute solution in which chemical reaction is very fast, from equation (2.14)

$$\frac{\partial C_A}{\partial \theta} = D_{AS}\left[\frac{\partial^2 C_A}{\partial z^2}\right] \quad for \quad 0 < z < z_R \qquad (5.164)$$

$$\frac{\partial C_B}{\partial \theta} = D_{BS}\left[\frac{\partial^2 C_B}{\partial z^2}\right] \quad for \quad z_R < z < \infty \qquad (5.165)$$

The distance of the reaction plane, z_R, from the interface, is a function of time. That is

$$z = f(\theta) \qquad (5.166)$$

For a very fast reaction,
$$C_A(z_R, \theta) = 0 \tag{5.167}$$
so that
$$\frac{\partial z_R}{\partial \theta} = -\frac{\left(\frac{\partial C_A}{\partial \theta}\right)_{z_R}}{\left(\frac{\partial C_A}{\partial z_R}\right)_\theta} \tag{5.168}$$

Equations (5.164) and (5.165), obviously, have the same solution as the penetration theory solution. That is
$$C_A = C_{A_0} + (C_{A_i} - C_{A_0})\,erfc\,\frac{z}{2\sqrt{\theta D_{AB}}} \tag{2.47}$$
which we can express as
$$\frac{C_A}{C_{A_i}} = A_1 + A_2\,erf\,\frac{z}{2\sqrt{\theta D_{AS}}} \tag{5.169}$$
$$\frac{C_B}{C_{B_\infty}} = A_3 + A_4\,erf\,\frac{z}{2\sqrt{\theta D_{BS}}} \tag{5.170}$$
where A_1, A_2, A_3 and A_4 are constants. From equation (5.169)
$$\frac{\partial C_A}{\partial \theta} = A_2 C_{Ai} \frac{2}{\sqrt{\pi}} e^{-\frac{z^2}{4\theta D_{AS}}} \cdot \frac{z}{2\sqrt{D_{AS}}}\left(-\frac{1}{2} \cdot \frac{1}{\theta^{\frac{3}{2}}}\right) \tag{5.171}$$

$$\frac{\partial C_A}{\partial z} = A_2 C_{Ai} \frac{2}{\sqrt{\pi}} e^{-\frac{z^2}{4\theta D_{AS}}} \cdot \frac{1}{2\sqrt{\theta D_{AS}}} \tag{5.172}$$

Substituting equations (5.171) and (5.172) into equation (5.168) and simplifying, we get that
$$\frac{d z_R}{d\theta} = \frac{z_R}{2\theta} \tag{5.173}$$
from which we get that
$$z = \frac{1}{2}\ln(\theta) + constant = 2\sqrt{A_5\,\theta}$$
or
$$\frac{z_R}{2\sqrt{D\theta}} = \sqrt{\frac{A_5}{D}} \tag{5.174}$$

CHAPTER FIVE: MASS TRANSFER WITH CHEMICAL REACTION

D may be D_{AS} or D_{BS}. We need the boundary conditions to evaluate the constants, A_1, A_2, A_3, A_4 and A_5, the initial conditions having been satisfied in the solution given by equation (5.169). The boundary conditions are

i	$z = 0$	$C_A = C_{Ai}$
ii	$z = z_R$	$C_A = 0$
iii	$z = \infty$	$C_B = C_{B\infty}$
iv	$z = z_R$	$C_B = 0$
v	$z = z_R$	$N_A + \dfrac{a}{b} N_B = 0$

Using boundary condition (i) in equation (5.169)
$$1 = A_1 + A_2 \times 0 \quad or \quad A_1 = 1 \qquad (5.175)$$
Using boundary condition (ii) in equation (5.169)
$$0 = 1 + A_2 \, erf \frac{z}{2\sqrt{\theta D_{AS}}} \quad and \quad \frac{z}{2\sqrt{\theta D_{AS}}} = \sqrt{\frac{A_5}{D_{AS}}}$$

That is $A_2 = -\dfrac{1}{erf\sqrt{\dfrac{A_5}{D_{AS}}}} \qquad (5.176)$

Using boundary condition (iii) in equation (5.170)
$$1 = A_3 + A_4 \quad or \quad A_3 = 1 - A_4 \qquad (5.177)$$
Using boundary condition (iv) in equations (5.170) and (5.177)
$$0 = 1 - A_4 + A_4 \, erf \frac{z}{2\sqrt{\theta D_{BS}}} = 1 - A_4 \, erfc \sqrt{\frac{A_5}{D_{BS}}}$$

That is $A_4 = \dfrac{1}{erfc\sqrt{\dfrac{A_5}{D_{BS}}}} \qquad (5.178)$

From equations (5.177) and (5.178)
$$A_3 = 1 - \frac{1}{erfc\sqrt{\dfrac{A_5}{D_{AS}}}} \qquad (5.179)$$

From boundary condition (v)

$$N_A + \frac{a}{b} N_B = 0$$

or
$$-\frac{b}{a} D_{AS} \frac{\partial C_A}{\partial z}\bigg|_{z=z_R} = D_{BS} \frac{\partial C_B}{\partial z}\bigg|_{z=z_R} \quad (5.180)$$

Applying equation (5.180) to equations (5.169) and (5.170)

$$\frac{b}{a} \frac{D_{AS}}{D_{BS}} \frac{\partial C_A}{\partial z}\bigg|_{z=z_R} = \frac{b}{a} \frac{D_{AS}}{D_{BS}} \frac{1}{erf\sqrt{\frac{A_5}{D_{AS}}}} C_{Ai} \frac{2}{\sqrt{\pi}} e^{-\frac{A_5}{D_{AS}}} \frac{1}{2\sqrt{\theta D_{AS}}}$$

$$(5.181)$$

$$\frac{\partial C_B}{\partial z}\bigg|_{z=z_R} = \frac{C_{B\infty}}{erfc\sqrt{\frac{A_5}{D_{BS}}}} \frac{2}{\sqrt{\pi}} e^{-\frac{A_5}{D_{BS}}} \frac{1}{2\sqrt{\theta D_{BS}}} \quad (5.182)$$

Combining equations (5.181) and (5.182) according to equation (5.180), we get that

$$\frac{b}{a} \frac{D_{AS}}{D_{BS}} \frac{C_{Ai}}{C_{B\infty}} \frac{erfc\sqrt{\frac{A_5}{D_{BS}}}}{erf\sqrt{\frac{A_5}{D_{AS}}}} = e^{A_5\left(\frac{1}{D_{AS}} - \frac{1}{D_{BS}}\right)} \quad (5.183)$$

Although equation (5.183) is not explicit in A_5, all the variables in it, except A_5, are known. Hence A_5 can be evaluated and then used to obtain A_2, A_3 and A_4 from equations (5.176), (5.179) and (5.178) respectively.

Concentration Profile

Substituting for A_1 and A_2 in equation (5.169)

$$\frac{C_A}{C_{A_i}} = 1 - \frac{erf\frac{z}{2\sqrt{\theta D_{AS}}}}{erf\sqrt{\frac{A_5}{D_{AS}}}} \quad \text{for} \quad 0 < z < z_R \quad (5.184)$$

Substituting for A_3 and A_4 in equation (5.170)

$$\frac{C_B}{C_{B\infty}} = 1 - \frac{1}{erfc\sqrt{\frac{A_5}{D_{BS}}}} + \frac{erf\frac{z}{2\sqrt{\theta D_{BS}}}}{erfc\sqrt{\frac{A_5}{D_{BS}}}}$$

$$= 1 - \frac{erfc\frac{z}{2\sqrt{\theta D_{BS}}}}{erfc\sqrt{\frac{A_5}{D_{BS}}}} \quad for \quad z_R < z < \infty \quad (5.185)$$

Instantaneous Molar Flux

$$N_{Ai}(\theta) = N_A(0,\theta) = -D_{AS}\frac{\partial C_A}{\partial z}\bigg|_{z=0}$$

$$= +D_{AS} C_{Ai} \cdot \frac{2}{\sqrt{\pi}} \cdot \frac{1}{2\sqrt{\theta D_{AS}}} \cdot \frac{1}{erf\sqrt{\frac{A_5}{D_{AS}}}}$$

$$= \sqrt{\frac{D_{AS}}{\pi \theta}} \cdot \frac{C_{Ai}}{erf\sqrt{\frac{A_5}{D_{AS}}}} \quad (5.186)$$

If we recall that, for physical absorption, the penetration theory gave the instantaneous mass transfer coefficient as

$$k_L(\theta) = \sqrt{\frac{D_{AS}}{\pi \theta}} \quad from \quad equation\ (2.51)$$

Equation (5.186) can, then, be expressed as

$$N_{Ai}(\theta) = k_L(\theta) \cdot \frac{C_{Ai}}{erf\sqrt{\frac{A_5}{D_{AS}}}} \quad 5.187)$$

Since $erf\sqrt{\frac{A_5}{D_{AS}}}$ is never greater than 1, equation (5.187) shows that the instantaneous molar flux for absorption with chemical reaction is greater than that for physical absorption.

Average Molar Flux

The average molar flux depends on the model of surface renewal used to evaluate it.

(a) **The Higbie Model**

τ is the contact time for all surface elements. Then

$$N_{Ai} = \frac{1}{\tau}\int_0^\tau N_{Ai}(\theta)d\theta = 2\sqrt{\frac{D_{AS}}{\pi\tau}}\frac{C_{Ai}}{erf\sqrt{\frac{A_5}{D_{AS}}}} \quad (5.188)$$

From equation (3.107) and (5.188)

$$N_{Ai} = k_L \frac{C_{Ai}}{erf\sqrt{\frac{A_5}{D_{AS}}}}$$

$$or \quad (k_L)_R = \frac{N_{Ai}}{C_{Ai}} = \frac{k_L}{erf\sqrt{\frac{A_5}{D_{AS}}}} \quad (5.189)$$

where $(k_L)_R$ is the Higbie model definition of the liquid phase mass transfer coefficient for unsteady state absorption.

(b) **Danckwert's Model**

$$N_{Ai} = \int_0^\infty \varphi(\theta) N_{Ai}(\theta) d\theta = \int_0^\infty se^{-s\theta}\sqrt{\frac{D_{AS}}{\pi\theta}}\frac{C_{Ai}}{erf\sqrt{\frac{A_5}{D_{AS}}}}d\theta$$

$$= \sqrt{D_{AS}\cdot s}\cdot \frac{C_{Ai}}{erf\sqrt{\frac{A_5}{D_{AS}}}} = k_L \frac{C_{Ai}}{erf\sqrt{\frac{A_5}{D_{AS}}}} \quad (5.190)$$

Here, $k_L = \sqrt{D_{AS}\cdot s}$ from equation (3.114).

CHAPTER FIVE: MASS TRANSFER WITH CHEMICAL REACTION

5.4.1.1: Gas Absorption with Very Fast Reaction with $D_{AS} = D_{BS} = D$

When $D_{AS} = D_{BS} = D$, equation (5.183) becomes

$$\frac{b}{a}\frac{D}{D}\frac{C_{Ai}}{C_{B\infty}}\frac{erfc\sqrt{\frac{A_5}{D}}}{erf\sqrt{\frac{A_5}{D}}} = e^{A_5\left(\frac{1}{D}-\frac{1}{D}\right)} = 1 \quad (5.191)$$

That is

$$\frac{b}{a}\frac{C_{Ai}}{C_{B\infty}}\frac{1-erf\sqrt{\frac{A_5}{D}}}{erf\sqrt{\frac{A_5}{D}}} = 1$$

or

$$\frac{b}{a}\frac{C_{Ai}}{C_{B\infty}}\left(\frac{1}{erf\sqrt{\frac{A_5}{D}}} - 1\right) = 1$$

Hence

$$\frac{1}{erf\sqrt{\frac{A_5}{D}}} = \frac{a}{b}\frac{C_{B\infty}}{C_{Ai}} + 1 \quad (5.192)$$

Concentration Profile

Equation (5.184) becomes, with equation (5.192)

$$\frac{C_A}{C_{Ai}} = 1 - \left(1 + \frac{a}{b}\frac{C_{B\infty}}{C_{Ai}}\right)erf\frac{z}{2\sqrt{\theta D}} \quad for \quad 0 < z < z_R \quad (5.193)$$

and equation (5.185) becomes

$$\frac{C_B}{C_{B\infty}} = 1 - \left(1 + \frac{b}{a}\frac{C_{Ai}}{C_{B\infty}}\right)erf\frac{z}{2\sqrt{\theta D}} \quad for \quad 0 < z < z_R \quad (5.194)$$

Instantaneous Molar Flux

$$N_{Ai}(\theta) = \sqrt{\frac{D}{\pi\theta}} \cdot \left(C_{Ai} + \frac{a}{b}C_{B\infty}\right) \quad (5.195)$$

Average Molar Flux

$$N_{Ai} = k_L \cdot \left(C_{Ai} + \frac{a}{b} C_{B\infty} \right) \quad (5.196)$$

where $k_L = \sqrt{D.s}$ for the Danckwert's model, and $k_L = \sqrt{\frac{D}{\pi \theta}}$ for the Higbie model, of surface renewal.

5.4.2: Gas Absorption with a First Order Chemical Reaction

It is assumed that the rates of molecular diffusion and reaction are comparable and that the penetration theory model applies. Then

$$\frac{\partial C_A}{\partial \theta} = D_{AS} \frac{\partial^2 C_A}{\partial z^2} - k_1 C_A \quad (5.197)$$

The initial and boundary conditions are

i $\theta = 0$ $0 < z < \infty$ $C_A = 0$
ii $\theta > 0$ $z = 0$ $C_A = C_{Ai}$, constant (5.198)
iii $\theta \geq 0$ $z = \infty$ $C_A = 0$

Applying the Laplace transform to equation (5.197)

$$p\overline{C_A} - C_A(0) = D_{AS} \frac{d^2 \overline{C_A}}{dz^2} - k_1 \overline{C_A} \quad (5.199)$$

and to the initial and boundary conditions (5.198)

i $\theta = 0$ $0 < z < \infty$ $\overline{C_A} = 0$
ii $\theta > 0$ $z = 0$ $\overline{C_A} = C_{Ai}/p$ (5.200)
iii $\theta \geq 0$ $z = \infty$ $\overline{C_A} = 0$

From initial condition (i) of equations (5.198) and equation (5.199)

$$\frac{d^2 \overline{C_A}}{dz^2} - \frac{k_1 + p}{D_{AS}} \overline{C_A} = 0 \quad (5.201)$$

Equation (5.201) has the solution, in the Laplace transform domain,

$$\overline{C_A} = A e^{\sqrt{\frac{k_1 + p}{D_{AS}}} \cdot z} + B e^{-\sqrt{\frac{k_1 + p}{D_{AS}}} \cdot z} \quad (5.202)$$

Using boundary condition (iii) of (5.200) in equation (5.202)
$$0 = A \times \infty + B \times 0 \quad \text{or} \quad A = 0 \quad (5.203)$$
Using boundary condition (ii) of (5.200) in equation (5.202)
$$\frac{\overline{C_{Ai}}}{p} = B \times 1 \quad \text{or} \quad B = \frac{\overline{C_{Ai}}}{p} \quad (5.204)$$
Substituting equations (5.203) and (5.204) into equation (5.202)
$$\overline{C_A} = \frac{\overline{C_{Ai}}}{p} e^{-\sqrt{\frac{k_1+p}{D_{AS}}} \cdot z} \quad (5.205)$$
which, from Laplace Transform Tables, gives
$$\frac{C_A}{C_{Ai}} = \frac{1}{2} e^{\sqrt{\frac{k_1}{D_{AS}}} \cdot z} \, erfc\left(\frac{z}{2\sqrt{\theta D_{AS}}} + \sqrt{k_1 \theta}\right)$$
$$+ \frac{1}{2} e^{-\sqrt{\frac{k_1}{D_{AS}}} \cdot z} \, erfc\left(\frac{z}{2\sqrt{\theta D_{AS}}} - \sqrt{k_1 \theta}\right) \quad (5.206)$$

<u>Instantaneous Molar Flux</u>

$$N_{Ai}(\theta) = N_A(0,\theta) = -D_{AS} \frac{\partial C_A}{\partial z}\bigg|_{z=0}$$

$$= -D_{AS} \frac{C_{Ai}}{2} \left[\begin{array}{l} \sqrt{\frac{k_1}{D_{AS}}} erfc\left(\sqrt{k_1 \theta}\right) + (-1)\frac{2}{\sqrt{\pi}} e^{-k_1 \theta} \frac{1}{2\sqrt{\theta D_{AS}}} \\ + (-)\sqrt{\frac{k_1}{D_{AS}}} erfc\left(-\sqrt{k_1 \theta}\right) + (-1)\frac{2}{\sqrt{\pi}} e^{-k_1 \theta} \frac{1}{2\sqrt{\theta D_{AS}}} \end{array}\right]$$

$$= D_{AS} \frac{C_{Ai}}{2} \left(\sqrt{\frac{k_1}{D_{AS}}} \left[erfc\left\{-\sqrt{k_1 \theta}\right\} - erfc\left\{\sqrt{k_1 \theta}\right\}\right] + \frac{2e^{-k_1 \theta}}{\sqrt{\theta \pi D_{AS}}}\right) \quad (5.207)$$

But

$$erfc\left\{-\sqrt{k_1 \theta}\right\} - erfc\left\{\sqrt{k_1 \theta}\right\} = erf\left\{-\sqrt{k_1 \theta}\right\} - erf\left\{\sqrt{k_1 \theta}\right\}$$

$$= \frac{2}{\sqrt{\pi}} \int_0^{\sqrt{k_1 \theta}} e^{-u^2} du - \frac{2}{\sqrt{\pi}} \int_0^{-\sqrt{k_1 \theta}} e^{-u^2} du$$

$$= \frac{2}{\sqrt{\pi}} \int_0^{\sqrt{k_1 \theta}} e^{-u^2} du + \frac{2}{\sqrt{\pi}} \int_{-\sqrt{k_1 \theta}}^0 e^{-u^2} du$$

$$= \frac{2}{\sqrt{\pi}} \int_{\sqrt{k_1\theta}}^{\sqrt{k_1\theta}} e^{-u^2} du = 2\,erf\left\{\sqrt{k_1\,\theta}\right\} \qquad (5.208)$$

Hence, from equations (5.208) and (5.207),

$$N_{Ai}(\theta) = C_{Ai}\left(\sqrt{k_1 D_{AS}}\cdot erf\left\{\sqrt{k_1\,\theta}\right\} + \sqrt{\frac{D_{AS}}{\pi\,\theta}}\,e^{-k_1\theta}\right) \qquad (5.209)$$

Since $k_L(\theta) = \sqrt{\dfrac{D}{\pi\,\theta}}$ for physical absorption, if we define

$$\beta = \frac{\sqrt{k_1 D_{AS}}}{k_L(\theta)} = \sqrt{\pi\,k_1\,\theta} \qquad (5.210)$$

then

$$N_{Ai}(\theta) = k_L(\theta) C_{Ai}\left(\beta(\theta)\cdot erf\left\{\frac{\beta(\theta)}{\sqrt{\pi}}\right\} + e^{-\frac{\beta^2(\theta)}{\pi}}\right) \qquad (5.211)$$

For a <u>fast reaction</u>, β is large [$\beta(\theta) \geq 3.0$] and

$$erf\left\{\frac{\beta(\theta)}{\sqrt{\pi}}\right\} \to 1.0, \quad \frac{\beta(\theta)}{\sqrt{\pi}} \geq 2.0 \text{ so that } e^{-\frac{\beta^2(\theta)}{\pi}} \to 0.$$

Then $N_{Ai}(\theta) = k_L(\theta) C_{Ai}\,\beta(\theta) = \sqrt{k_1 D_{AS}}\cdot C_{Ai}$ (5.212)

That is, the instantaneous, interfacial, molar flux is independent of time.

For a <u>slow reaction</u>, β is small [$\beta(\theta) \leq 0.35$] and

$$erf\left\{\frac{\beta(\theta)}{\sqrt{\pi}}\right\} = \frac{2}{\sqrt{\pi}}\frac{\beta(\theta)}{\sqrt{\pi}} = \frac{2\beta(\theta)}{\pi}, \text{ that is } \frac{\beta(\theta)}{\sqrt{\pi}} \leq 0.2.$$

$$e^{-\frac{\beta^2(\theta)}{\pi}} = 1 - \frac{\beta^2(\theta)}{\pi} \qquad (5.213)$$

Substituting equation (5.213) into equation (5.211)

$$N_{Ai}(\theta) = k_L(\theta) C_{Ai}\left(\frac{2\beta^2(\theta)}{\pi} + 1 - \frac{\beta^2(\theta)}{\pi}\right)$$

$$= k_L(\theta) C_{Ai}\left(1 + \frac{\beta^2(\theta)}{\pi}\right)$$

$$= k_L(\theta) C_{Ai}\left(1 + k_1\,\theta\right) \qquad (5.214)$$

Average Molar Flux

(a) The Higbie Model

τ is the contact time for all surface elements. Then

$$N_{Ai} = \frac{1}{\tau} \cdot \int_0^\tau N_{Ai}(\theta) d\theta$$

$$= \frac{1}{\tau} \cdot \int_0^\tau k_L(\theta) C_{Ai} \left(\beta(\theta).erf\left\{\frac{\beta(\theta)}{\sqrt{\pi}}\right\} + e^{-\frac{\beta^2(\theta)}{\pi}} \right) d\theta$$

$$N_{Ai} = k_L C_{Ai} \left(\left[\beta + \frac{\pi}{8\beta}\right].erf\left\{\frac{2\beta}{\sqrt{\pi}}\right\} + \frac{1}{2} e^{-\frac{4\beta^2}{\pi}} \right) \quad (5.215)$$

Note that

$$\int x\, erf\, x\, dx = \frac{1}{2} x^2 erf\, x - \frac{1}{\sqrt{\pi}} \int x^2 e^{-x^2} dx \quad (5.216)$$

$$\int x^2 e^{-x^2} dx = -\frac{1}{2} x e^{-x^2} + \frac{1}{2} \int e^{-x^2} dx \quad (5.217)$$

For a <u>fast reaction</u>, β is large. Then

$$N_{Ai} = k_L C_{Ai} \beta = \sqrt{k_1 D_{AS}} . C_{Ai} \quad (5.218)$$

For a <u>slow reaction</u>, β is small

$$N_{Ai} = k_L C_{Ai} \left(\left[\beta + \frac{\pi}{8\beta}\right].\frac{2}{\sqrt{\pi}}.\frac{2\beta}{\sqrt{\pi}} + \frac{1}{2}\left[1 - \frac{4\beta^2}{\pi}\right] \right) \quad (5.219)$$

$$= k_L C_{Ai} \left(\frac{4\beta^2}{\pi} + \frac{1}{2} + \frac{1}{2} - \frac{2\beta^2}{\pi} \right)$$

$$= k_L C_{Ai} \left(1 + \frac{2\beta^2}{\pi}\right) = k_L C_{Ai} \left(1 + \frac{1}{2} k_1 \tau\right) \quad (5.220)$$

(b) Danckwert's Model

$$N_{Ai} = \int_0^\infty \varphi(\theta) N_{Ai}(\theta) d\theta = s \int_0^\infty e^{-s\theta} N_{Ai}(\theta) d\theta$$

$$\overline{sN_{Ai}(\theta)} = s\overline{N_{Ai}(\theta)} \qquad (5.221)$$

where $\overline{N_{Ai}(\theta)}$ is the Laplace transform of $N_{Ai}(\theta)$.
From equation (5.209)

$$N_{Ai}(\theta) = C_{Ai}\left(\sqrt{k_1 D_{AS}} \cdot erf\{\sqrt{k_1\theta}\} + \sqrt{\frac{D_{AS}}{\pi\theta}} e^{-k_1\theta}\right) \qquad (5.209)$$

Therefore

$$s.\overline{N_{Ai}(\theta)} = C_{Ai}\left(\sqrt{k_1 D_{AS}} \cdot \frac{\sqrt{k_1}}{\sqrt{k_1+s}} + s.\sqrt{\frac{D_{AS}}{\pi}}\frac{\sqrt{\pi}}{\sqrt{k_1+s}}\right) \qquad (5.222)$$

$$s.\overline{N_{Ai}(\theta)} = C_{Ai}\left(\frac{k_1+s}{\sqrt{k_1+s}}\right)\sqrt{D_{AS}} = C_{Ai}\sqrt{D_{AS}(k_1+s)} \qquad (5.223)$$

Note that, from Laplace Transform Tables, the Laplace transform of the terms in equation (5.209) would be given as

$$\overline{erf\sqrt{k_1\theta}} = \frac{1}{s}\frac{\sqrt{k_1}}{\sqrt{k_1+s}} \qquad (5.224)$$

$$\overline{\left(\theta^{-n}e^{-k_1\theta}\right)} = \Gamma(n)(k_1+s)^{-n} \qquad (5.225)$$

where $\Gamma(n)$ is the gamma function. Thus, from equation (5.225)

$$\overline{\left(\theta^{-1/2}e^{-k_1\theta}\right)} = \Gamma(1/2)\frac{1}{\sqrt{k_1+s}} = \frac{\sqrt{\pi}}{\sqrt{k_1+s}} \qquad (5.226)$$

The gamma function is given by

$$\Gamma(n) = \int x^{n-1}e^{-x}dx \qquad (5.227)$$

$$\Gamma(1/2) = \int_0^\infty \frac{e^{-x}}{\sqrt{x}}dx = 2\int_0^\infty e^{-x}dx$$

$$= \frac{1}{2}\sqrt{\pi}.2 = \sqrt{\pi} \qquad (5.228)$$

Thus, from equations (5.221) and (5.223)

$$\overline{N_{Ai}} = s.\overline{N_{Ai}(\theta)} = C_{Ai}\sqrt{D_{AS}(k_1+s)} \qquad (5.229)$$

For physical absorption, the Danckwert's model gave $k_L = \sqrt{D_{AS}.s}$ [from equation (3.114)]. That means that

$$\beta = \frac{\sqrt{k_1 D_{AS}}}{k_L} = \sqrt{\frac{k_1}{s}} \qquad (5.230)$$

CHAPTER FIVE: MASS TRANSFER WITH CHEMICAL REACTION

From equations (5.229) and (5.230)
$$N_{Ai} = k_L C_{Ai}\sqrt{1 + \frac{k_1}{s}} = k_L C_{Ai}\sqrt{1 + \beta^2} \qquad (5.231)$$

For a <u>fast reaction</u>, β is large. Then $\beta^2 \gg 1$; $k_1 \gg s$
$$N_{Ai} = k_L C_{Ai} \beta = \sqrt{k_1 D_{AS}} \cdot C_{Ai} \qquad (5.232)$$

For a <u>slow reaction</u>, β is large. Then $\beta^2 \ll 1$; $k_1 \ll s$
$$N_{Ai} = k_L C_{Ai} \qquad (5.233)$$

5.5: Mass Transfer with Chemical Reaction in Heterogeneous Medium

5.5.1: Instantaneous Reaction at a Catalyst Surface

Consider a polymerisation reaction, for example
$$aA = A_a \qquad (5.234)$$
where A is the monomer and $A_a = p =$ the polymer. If we define $N_P =$ the molar flux of polymer = rate of formation of polymer per unit catalyst surface, then, for a binary system, consisting only of monomer and polymer,
$$N_P = (N_A + N_P)y_P + n_P \qquad (5.235)$$
From stoichiometry,
$$N_A = -aN_P \qquad (5.236)$$
Hence
$$N_P = \frac{n_P}{1 + (a-1)y_P} = -\frac{DC_T}{1 + (a-1)y_P}\frac{dy_P}{dz} \qquad (5.237)$$
On integration, with $y_P(0) = 1.0$ and $y_P(L) = y_{Pb}$
$$N_P = \frac{DC_T}{(a-1)L}\ln\frac{a}{1 + (a-1)y_{Pb}}$$
$$= \frac{(k_G)_y}{(a-1)}\ln\frac{a}{1 + (a-1)y_{Pb}}$$
$$= \frac{\pi(k_G)_P}{(a-1)}\ln\frac{a\pi}{1 + (a-1)P_{Pb}} \qquad (5.238)$$

For dimerisation, for example, $a = 2$ and equation (5.238) becomes

$$N_{P2} = \pi(k_G)_P \ln\frac{2\pi}{1+P_{Pb}} \qquad (5.239)$$

5.5.2: Diffusion and Chemical Reaction in Catalyst Particles

Consider an element of volume of a catalyst particle, with steady state and isothermal conditions within the particle, shown below. Uni-directional mass transfer occurs along z-direction as shown.

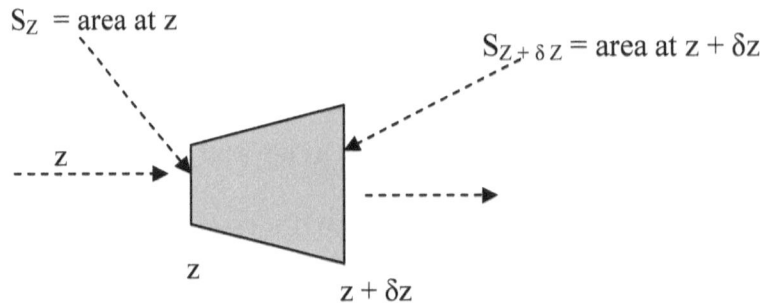

A steady state mass balance based on

Input + Generation = output + accummulation (5.240)

gives
$$(N_{Az}S)_z + R_A\delta V = (N_{Az}S)_{z+\delta z} \qquad (5.241)$$
where R_A is the rate of formation per unit volume (pores included) of catalyst pellet.

From equation (5.241)
$$R_A\delta V = (N_{Az}S)_z + \frac{d(N_{Az}S)_z}{dz}\delta z - (N_{Az}S)_z \qquad (5.242)$$

Hence
$$R_A = \frac{1}{S}\frac{d(N_{Az}S)_z}{dz} = \frac{dN_{Az}}{dz} + \frac{N_{Az}}{S}\frac{dS}{dz} \qquad (5.243)$$

But
$$N_{Az} = -D_A\frac{dC_A}{dz} \qquad (5.244)$$

CHAPTER FIVE: MASS TRANSFER WITH CHEMICAL REACTION

where D_A is an experimental value of a pore diffusion coefficient.

Thus, from equations (5.244) and (5.243), we get the general expression

$$R_A = -D_A \frac{d^2 C_A}{dz^2} - \frac{D_A}{S} \frac{dS}{dz} \frac{dC_A}{dz} \quad (5.245)$$

For the three types of catalyst particles in use, we get that for
(i) <u>Flat Particles</u>

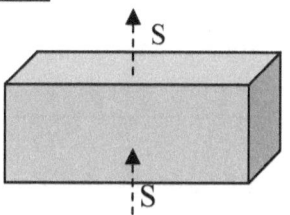

S is constant so that $\dfrac{dS}{dz} = 0$ (5.246)

And $R_A = -D_A \dfrac{d^2 C_A}{dz^2}$ (5.247)

(ii) <u>Cylindrical Particles</u>

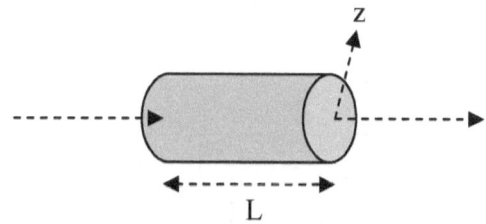

$S = 2\pi z L$ so that $\dfrac{dS}{dz} = 2\pi L$ (5.248)

Thus, from equations (5.248) and (5.245)

$$R_A = -D_A \frac{d^2 C_A}{dz^2} - \frac{D_A}{2\pi z L} 2\pi L \frac{dC_A}{dz}$$

$$= -D_A \frac{d^2 C_A}{dz^2} - \frac{D_A}{z} \frac{dC_A}{dz} \quad (5.249)$$

(iii) <u>Spherical Particles</u>

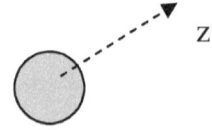

$$S = 4\pi z^2 \quad \text{so that} \quad \frac{dS}{dz} = 8\pi z \quad (5.250)$$

Thus, from equations (5.250) and (5.245)

$$R_A = -D_A \frac{d^2 C_A}{dz^2} - \frac{D_A}{4\pi z^2} 8\pi z \frac{dC_A}{dz}$$

$$= -D_A \frac{d^2 C_A}{dz^2} - \frac{2D_A}{z} \frac{dC_A}{dz} \quad (5.251)$$

Boundary conditions, applicable to the three forms of catalyst particles, are

i $\quad z = 0 \quad \dfrac{dC_A}{dz} = 0 \quad (5.252)$

ii $\quad z = b \quad C_A = C_{Ai}$

where b is the distance, in the z-direction, from a point or axis of symmetry to the surface of the particle.

5.5.3: Steady State Diffusion with Chemical Reaction in Flat Catalyst Particles

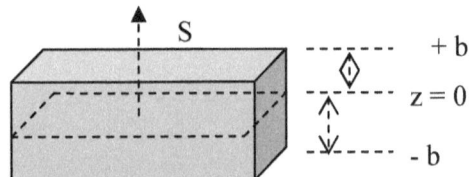

For the system shown above, the relevant equation to be solved is, from equation (5.247),

$$\frac{d^2 C_A}{dz^2} + \frac{R_A}{D_A} = 0 \quad (5.247a)$$

with boundary conditions

i $\quad z = 0 \quad \dfrac{dC_A}{dz} = 0 \quad (5.252)$

CHAPTER FIVE: MASS TRANSFER WITH CHEMICAL REACTION

ii $z = b$ $C_A = C_{Ai}$

5.5.3.1: Zero Order Reaction

$$R_A = -k_0 \quad (5.253)$$

and

$$\frac{d^2 C_A}{dz^2} = \frac{k_0}{D_A} \quad \text{or} \quad \frac{dC_A}{dz} = \frac{k_0}{D_A} z + A$$

which gives

$$C_A = \frac{k_0}{2 D_A} z^2 + Az + B \quad (5.254)$$

where A and B are integration constants. From boundary condition (i), $A = 0$. From boundary condition (ii)

$$B = C_{Ai} - \frac{k_0 b^2}{2 D_A} \quad (5.255)$$

Substituting for the values of A and B in equation (5.254)

$$C_A = C_{Ai} - \frac{k_0 b^2}{2 D_A} + \frac{k_0}{2 D_A} z^2 \quad (5.256)$$

a parabola.

Molar Flux

$$N_{Az} = -D_A \frac{dC_A}{dz}\bigg|_{z=b} = D_A \frac{k_0}{D_A} b$$

$$= k_0 b \quad (5.257)$$

That is, the molar flux depends only on the volume of the catalyst particle.

Effectiveness of Catalyst

It is usual to define the effectiveness of a catalyst as

Effectiveness of catalyst $= \varepsilon_A$

$$= \frac{\textit{Actual Rate of Reaction}}{\textit{Reaction Rate if all surface at bulk concentration}}$$

(5.258)

Thus, for a flat particle, cross sectional area, S, thickness, $2b$,

$$\varepsilon_A = \frac{N_{Ai} S_T}{R_{Ai} V_T} = \frac{2 S N_{Ai}}{2 b S R_{Ai}} = \frac{N_{Ai}}{b R_{Ai}} \quad (5.259)$$

For a zero order reaction

$$\varepsilon_A = \frac{-k_0 b}{b(-k_0)} = 1 \quad (5.260)$$

5.5.3.2: First Order Irreversible Reaction

$$R_A = -k_1 C_A \quad (5.261)$$

$$\frac{d^2 C_A}{dz^2} + \frac{k_1 C_A}{D_A} = 0 \quad (5.247b)$$

with boundary conditions

i $\quad z = 0 \quad \dfrac{dC_A}{dz} = 0 \quad (5.252)$

ii $\quad z = b \quad C_A = C_{Ai}$

Equation (5.247b) becomes, on integration,

$$C_A = A \exp\left(\sqrt{\frac{k_1}{D_A}} z\right) + B \exp\left(-\sqrt{\frac{k_1}{D_A}} z\right) \quad (5.262)$$

Applying boundary condition (i)

$$\left.\frac{dC_A}{dz}\right|_{z=0} = \sqrt{\frac{k_1}{D_A}} (A - B) = 0$$

That is $A = B \quad (5.263)$

Applying boundary condition (ii)

$$C_{Ai} = 2A \cosh\left(\sqrt{\frac{k_1}{D_A}} b\right)$$

Therefore $A = B = \dfrac{C_{Ai}}{2 \cosh\left(\sqrt{\dfrac{k_1}{D_A}} b\right)} \quad (5.264)$

and
$$C_A = C_{Ai} \frac{\cosh\left(\sqrt{\frac{k_1}{D_A}} z\right)}{2\cosh\left(\sqrt{\frac{k_1}{D_A}} b\right)} \qquad (5.265)$$

Molar Flux

$$N_{Ai} = -D_A \frac{dC_A}{dz}\bigg|_{z=b} = -D_A C_{Ai} \sqrt{\frac{k_1}{D_A}} \tanh\left(\sqrt{\frac{k_1}{D_A}} b\right)$$

$$= \sqrt{k_1 D_A} \cdot C_{Ai} \tanh\left(\sqrt{\frac{k_1}{D_A}} b\right) \qquad (5.266)$$

Catalyst Effectiveness

$$\varepsilon_A = \frac{\int_{V_T} R_A \, dV}{R_{Ai} V_T} = \frac{N_{Ai} S_T}{R_{Ai} V_T}$$

$$= \frac{-\sqrt{k_1 D_A} \cdot C_{Ai} \tanh\left(\sqrt{\frac{k_1}{D_A}} b\right) \cdot 2 \cdot S}{-k_1 C_{Ai} \cdot 2 \cdot b \cdot S}$$

$$= \frac{\sqrt{k_1 D_A}}{k_1 \cdot b} \tanh\left(\sqrt{\frac{k_1}{D_A}} b\right) = \frac{\tanh\left(\sqrt{\frac{k_1}{D_A}} b\right)}{\sqrt{\frac{k_1}{D_A}} b} \qquad (5.267)$$

For a Fast Reaction

Let $x = \sqrt{\frac{k_1}{D_A}} b$. When x is large

$$\tanh(x) = \frac{e^x - e^{-x}}{e^x + e^{-x}} = 1 \qquad (5.268)$$

and

$$\varepsilon_A = \frac{\tanh x}{x} = \frac{1}{\sqrt{\frac{k_1}{D_A}b}} \qquad (5.269)$$

That is, $\varepsilon_A \to \varepsilon_{A\max}$ as $b \to b_{\min}$. This means that, for a given, k_1 and D_A, b must be as small as possible.

For a Slow Reaction

$$x = \sqrt{\frac{k_1}{D_A}b} \text{ and } x \text{ is small}$$

$$\tanh(x) = \frac{1+x-1+x}{1+x+1-x} = x \qquad (5.270)$$

and

$$\varepsilon_A = \frac{\tanh x}{x} = 1 \qquad (5.271)$$

That is, concentration of reactant in the pores of the catalyst is, approximately, equal to its concentration outside the particle. This implies that molecular diffusion is greater than chemical reaction. These results may be illustrated in the diagram below.

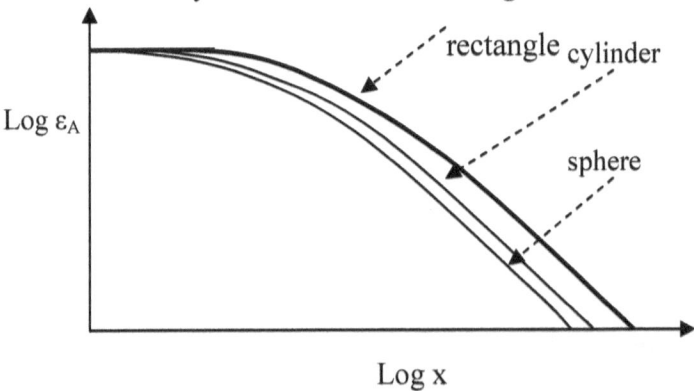

5.5.3.3: First Order Reversible Reaction

Consider a reversible first order reaction such as that represented below

CHAPTER FIVE: MASS TRANSFER WITH CHEMICAL REACTION

$$aA + M \underset{k_B}{\overset{k_A}{\rightleftarrows}} bB + N \qquad (5.272)$$

It is assumed that this reaction is first order with regard to A and B in the forward and reverse directions, respectively.

The rate equation is
$$R_A = -k_A C_A + k_B C_B \qquad (5.273)$$
while the material balance equation is
$$D_A \frac{d^2 C_A}{dz^2} + R_A = 0 \qquad (5.247c)$$

From stoichiometry,

$$\frac{N_A}{N_B} = -\frac{a}{b} = \frac{-D_A \dfrac{dC_A}{dz}}{-D_B \dfrac{dC_B}{dz}} \qquad (5.274)$$

from which we get that

$$\frac{D_A}{D_B} \cdot \frac{b}{a} dC_A = -dC_B$$

or $\quad \dfrac{D_A}{D_B} \cdot \dfrac{b}{a} C_A + C_B = $ constant $\qquad (5.275)$

Similarly,

or $\quad \dfrac{D_A}{D_B} \cdot \dfrac{b}{a} C_{Ai} + C_{Bi} = $ constant $\qquad (5.276)$

Assuming the same constant for equations (5.275) and (5.276), then

$$C_B = C_{Bi} + \frac{D_A}{D_B} \cdot \frac{b}{a} \cdot (C_{Ai} - C_A) \qquad (5.277)$$

Substituting for C_B from equation (5.277) in equation (5.273)

$$R_A = -k_A C_A + k_B \left(C_{Bi} + \frac{D_A}{D_B} \cdot \frac{b}{a} \cdot (C_{Ai} - C_A) \right)$$

$$= -\left(k_A + k_B \frac{D_A}{D_B} \frac{b}{a} \right) C_A + k_B \left(C_{Bi} + \frac{D_A}{D_B} \frac{b}{a} C_{Ai} \right) \qquad (5.278)$$

At equilibrium

$$R_A = 0 \quad \text{and} \quad C_A = C_{Ae} \qquad (5.279)$$

so that

$$\left(k_A + k_B \frac{D_A}{D_B} \frac{b}{a}\right) C_{Ae} = k_B \left(C_{Bi} + \frac{D_A}{D_B} \frac{b}{a} C_{Ai}\right) \qquad (5.280)$$

From equations (5.280) and (5.278)

$$R_A = -\left(k_A + k_B \frac{D_A}{D_B} \frac{b}{a}\right)(C_A - C_{Ae})$$

$$= D_A \alpha^2 (C_A - C_{Ae}) \qquad (5.281)$$

where

$$\alpha^2 = \left(\frac{k_A}{D_A} + \frac{k_B}{D_B} \frac{b}{a}\right) \qquad (5.282)$$

Substituting equation (5.281) into equation (5.247c)

$$\frac{d^2 C_A}{dz^2} - \alpha^2 (C_A - C_{Ae}) = 0 \qquad (5.283)$$

The boundary conditions are

i $\quad z = 0 \qquad \dfrac{d(C_A - C_{Ae})}{dz} = 0 \qquad (5.284)$

ii $\quad z = b \qquad C_A = C_{Ai}$

This enables the solution of equation (5.283) as

$$\frac{C_A - C_{Ae}}{C_{Ai} - C_{Ae}} = \frac{\cosh(\alpha z)}{\cosh(\alpha b)} \qquad (5.285)$$

Molar Flux

$$N_{Ai} = -D_A \frac{dC_A}{dz}\bigg|_{z=b}$$

$$= -D_A \alpha (C_{Ai} - C_{Ae}) + \tanh(\alpha b) \qquad (5.286)$$

Catalyst Effectiveness

$$\varepsilon_A = \frac{[-D_A \alpha (C_{Ai} - C_{Ae}) + \tanh(\alpha b)].2.S}{\left(-D_A \alpha^2 (C_{Ai} - C_{Ae})\right)2.b.S}$$

$$= \frac{\tanh(\alpha b)}{\alpha b} \qquad (5.287)$$

CHAPTER FIVE: MASS TRANSFER WITH CHEMICAL REACTION

For a fast reaction
$$\tanh(\alpha b) \to 1 \quad \text{and} \quad \varepsilon_A \to \frac{1}{qb} \qquad (5.288)$$

For a slow reaction
$$\tanh(\alpha b) \to qb \quad \text{and} \quad \varepsilon_A \to 1 \qquad (5.289)$$

5.5.4: Steady State Diffusion with Chemical Reaction in Cylindrical Catalyst Particles

First Order Irreversible Reaction

$$R_A = -k_1 C_A \qquad (5.261)$$

For cylindrical catalyst particles, the material balance is

$$R_A = -D_A \frac{d^2 C_A}{dz^2} - \frac{D_A}{z} \frac{dC_A}{dz} \qquad (5.249)$$

From equations (5.261) and (5.249)

$$\frac{d^2 C_A}{dz^2} + \frac{1}{z} \frac{dC_A}{dz} - q^2 C_A = 0 \qquad (5.290)$$

with boundary conditions (5.252). $q^2 = \frac{k_1}{D_A}$. The solution is

$$C_A = A I_0(qz) + B K_0(qz) \qquad (5.291)$$

A and B are integration constants and I is the Bessel's function.
From boundary condition (i)
$$K_0(0) = \infty$$
that is $B = 0$ if C_A is finite $\qquad (5.292)$

From boundary condition (ii)

$$C_{Ai} = A I_0(qb) \quad \text{or} \quad A = \frac{C_{Ai}}{I_0(qb)} \qquad (5.293)$$

Thus, equation (5.291) becomes

$$C_A = C_{Ai} \cdot \frac{I_0(qz)}{I_0(qb)} \qquad (5.294)$$

Molar Flux

$$N_{Ai} = -D_A \frac{dC_A}{dz} \bigg|_{z=b}$$

$$= -D_A \frac{qC_{Ai}}{I_0(qb)} \frac{dI_0(qz)}{dz} = -D_A qC_{Ai} \frac{I_1(qb)}{I_0(qb)}$$

$$= -\sqrt{D_A k_1} C_{Ai} \frac{I_1(qb)}{I_0(qb)} \qquad (5.295)$$

Catalyst Effectiveness

$$\varepsilon_A = \frac{N_{Ai} S_T}{R_{Ai} V_T} = \frac{2\pi b L N_{Ai}}{\pi b^2 L R_{Ai}}$$

$$= \frac{2 N_{Ai}}{b R_{Ai}} = \frac{2}{qb} \cdot \frac{I_1(qb)}{I_0(qb)} \qquad (5.296)$$

For a fast reaction

$$\frac{I_1(qb)}{I_0(qb)} \to 1 \quad and \quad \varepsilon_A \to \frac{1}{qb} \qquad (5.297)$$

5.5.5: Steady State Diffusion with Chemical Reaction in Spherical Catalyst Particles

First Order Irreversible Reaction

$$R_A = -k_1 C_A \qquad (5.261)$$

For spherical catalyst particles, the material balance is, from equation (5.251) and (5.261),

$$z \frac{d^2 C_A}{dz^2} + 2 \frac{dC_A}{dz} - \frac{k_1 z}{D_A} C_A = 0$$

or $\quad \dfrac{1}{z} \dfrac{d}{dz}\left(z^2 \dfrac{dC_A}{dz}\right) - \dfrac{k_1 z}{D_A} C_A = 0 \quad (5.298)$

The boundary conditions are

i $\quad z = 0 \qquad \dfrac{dC_A}{dz} = 0 \qquad (5.252)$

ii $\quad z = b \qquad C_A = C_{Ai}$

where b is the distance, in the z-direction, from the centre of the sphere to the surface of the particle.

CHAPTER FIVE: MASS TRANSFER WITH CHEMICAL REACTION

If we let
$$C_A = \frac{f(z)}{z} \qquad (5.299)$$

Then
$$\frac{1}{z}\frac{d}{dz}\left(z^2 \frac{dC_A}{dz}\right) = z\frac{d^2 f(z)}{dz^2} \qquad (5.300)$$

Equations (5.300) and (5.299) enable us to express equation (5.298) as

$$\frac{d^2 f(z)}{dz^2} - q^2 f(z) = 0 \qquad (5.301)$$

where $q^2 = \frac{k_1}{D_A}$. The solution is

$$f(z) = A\exp(qz) + B\exp(-qz) \qquad (5.302)$$

The boundary conditions (5.252) become, from equation (5.296)
I $z = 0$ C_A is finite $f(z) = 0$ (5.252)
Ii $z = b$ $C_A = C_{Ai}$ $f(b) = bC_{Ai}$

Applying boundary condition (i) to equation (5.302)
$$0 = A + B; \text{ that is } f(z) = 2A\sinh(qz) \qquad (5.303)$$

Applying boundary condition (ii) to equation (5.303)
$$2A = \frac{bC_{Ai}}{\sinh(qb)}; \text{ i.e. } f(z) = bC_{Ai}\frac{\sinh(qz)}{\sinh(qb)} \qquad (5.304)$$

From equations (5.304) and (5.299)

$$C_A = \frac{b}{z}C_{Ai}\frac{\sinh(qz)}{\sinh(qb)} \qquad (5.305)$$

Molar Flux

$$N_{Ai} = -D_A\frac{dC_A}{dz}\bigg|_{z=b}$$

$$= \frac{D_A}{b}C_{Ai}[1 - qb\coth(qb)] \qquad (5.306)$$

Catalyst Effectiveness

$$\varepsilon_A = \frac{N_{Ai} S_T}{R_{Ai} V_T} = \frac{4\pi b^2 N_{Ai}}{\frac{4}{3}\pi b^3 R_{Ai}} = \frac{3 N_{Ai}}{b R_{Ai}} \qquad (5.307)$$

which gives, from equations (5.307) and (5.306)

$$\varepsilon_A = \frac{3\dfrac{D_A}{b} C_{Ai}\left[1 - qb\coth(qb)\right]}{-bk_1 C_{Ai}}$$

$$= \frac{3 D_A}{b^2 k_1}\left[qb\coth(qb) - 1\right]$$

$$= \frac{3}{b^2 q^2}\left[qb\coth(qb) - 1\right] \qquad (5.308)$$

If $\lambda = b/3$

$$\varepsilon_A = \frac{1}{3(q\lambda)^2}\left[3(q\lambda)\coth(3q\lambda) - 1\right] \qquad (5.309)$$

For a fast reaction

$$\coth(3q\lambda) \to 1 \quad and \quad \varepsilon_A \to \frac{1}{q\lambda} \qquad (5.310)$$

5.5.6: Calculation of Isothermal Fixed Bed Reactors

The basic assumptions are
    (i)   the reactor is isothermal
    (ii)   pressure drop in the reactor is less than then total pressure in the system
    (iii) there is constant molal flow

For a packed bed, with void fraction, ϕ, the rate of formation, R_A^*, of product, A, per unit volume of reactor bed is given by

$$R_A^* = R_{Ai}(1 - \phi)\varepsilon_A \qquad (5.311)$$

A differential element of the packed bed may be illustrated as

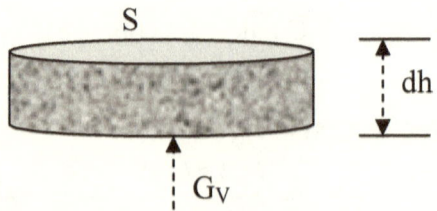

A material balance gives, assuming gas stream is in plug flow,

$$R_A^* S\,dh = N_{Ai}\, a_P\, S\, dh = G_V\, dC_A \qquad (5.312)$$

From equations (5.311) and (5.312)

CHAPTER FIVE: MASS TRANSFER WITH CHEMICAL REACTION

$$R_{Ai}(1 - \phi)\varepsilon_A \, dh = N_{Ai} a_P \, dh = u_G \, dC_A \quad (5.313)$$

where u_G is the superficial gas velocity $= G_V/S$. Thus from equation (5.313)

$$R_{Ai}(1 - \phi)\varepsilon_A = N_{Ai} a_P \quad (5.314)$$

and

$$z = \frac{u_G}{(1 - \phi)\varepsilon_A} \int \frac{dC_A}{R_{Ai}} = \frac{u_G}{a_P} \int \frac{dC_A}{N_{Ai}} \quad (5.315)$$

5.5.6.1: Zero Order Reaction

$$R_A^* = R_{Ai} = -k_0; \quad \varepsilon_A = 1 \quad (5.316)$$

$$z = \frac{u_G}{(1 - \phi)k_0} \int dC_A$$

$$= \frac{u_G}{(1 - \phi)k_0}(C_{A1} - C_{A2}) \quad (5.317)$$

5.5.6.2: First Order Irreversible Reaction

$$R_{Ai} = -k_1 C_{Ai}; \quad N_{Ai} = k_G(C_{Ai} - C_{Ab}) \quad (5.318)$$

From equations (5.314) and (5.318)

$$-k_1 C_{Ai}(1 - \phi)\varepsilon_A = k_G(C_{Ai} - C_{Ab})a_P \quad (5.319)$$

from which we get that

$$C_{Ai} = \frac{k_G a_P C_{Ab}}{k_G a_P + k_1(1 - \phi)\varepsilon_A} \quad (5.320)$$

and

$$R_{Ai} = -\frac{k_1 k_G a_P C_{Ab}}{k_G a_P + k_1(1 - \phi)\varepsilon_A} \quad (5.321)$$

From equations (5.321) and (5.315)

$$z = u_G \frac{k_G a_P + k_1(1 - \phi)\varepsilon_A}{k_1 k_G a_P (1 - \phi)\varepsilon_A} \int \frac{dC_{Ab}}{C_{Ab}} \quad (5.322)$$

Note that

$$z = (H_{OR})_1 \times (N_R)_1 \quad (5.323)$$

where

$$(N_R)_1 = \int \frac{dC_{Ab}}{C_{Ab}} = \ln \frac{C_{A1}}{C_{A2}} \quad (5.324)$$

and

$$(H_{OR})_1 = \frac{u_G}{k_1(1-\phi)\varepsilon_A} + \frac{u_G}{k_G a_P}$$
$$= (H_R)_1 + H_G \quad (5.325)$$

For a fast reaction

$$\varepsilon_A = \frac{1}{q\lambda} = \frac{1}{\lambda}\sqrt{\frac{D_A}{k_1}} \quad (5.326)$$

and

$$(H_R)_1 = \frac{u_G \lambda}{(1-\phi)\sqrt{k_1 D_A}} \quad (5.327)$$

5.5.6.3: First Order Reversible Reaction

$$R_{Ai} = -k_A C_{Ai} + k_B C_{Bi};$$
$$N_{Ai} = k_G(C_{Ai} - C_{Ab}) \quad (5.328)$$

For a binary mixture

$$C_{Bi} = C_T - C_{Ai} \quad \text{so that}$$
$$R_{Ai} = -(k_A + k_B)C_{Ai} + k_B C_T \quad (5.329)$$

From equations (5.329), (5.328) and (5.314)

$$C_{Ai} = \frac{k_G a_P C_{Ab} + k_B C_T \varepsilon_A (1-\phi)}{k_G a_P + \varepsilon_A (1-\phi)(k_A + k_B)} \quad (5.330)$$

and

$$R_{Ai} = -\frac{k_1 k_G a_P C_{Ab}}{k_G a_P + k_1(1-\phi)\varepsilon_A} \quad (5.331)$$

From equations (5.330) and (5.328)

$$N_{Ai} = \frac{k_G \varepsilon_A (1-\phi)(k_A + k_B)}{k_G a_P + \varepsilon_A (1-\phi)(k_A + k_B)}\left(\frac{k_B C_T}{k_A + k_B} - C_{Ab}\right) \quad (5.332)$$

From equations (5.332) and (5.315)

$$z = u_G \frac{k_G a_P + \varepsilon_A (1 - \phi)(k_A + k_B)}{k_G a_P (1 - \phi) \varepsilon_A (k_A + k_B)} \int_2^1 \frac{dC_{Ab}}{C_{Ab} - \frac{k_B}{k_A + k_B} C_T} \quad (5.333)$$

$$= (H_{OR})_{R1} \times (N_R)_{R1} \quad (5.334)$$

where

$$(H_{OR})_{R1} = \frac{u_G}{\varepsilon_A (1 - \phi)(k_A + k_B)} + \frac{u_G}{k_G a_P}$$

$$= (H_R)_1 + H_G \quad (5.335)$$

and

$$(N_R)_{R1} = \ln \frac{C_{A1} - \frac{k_B}{k_A + k_B} C_T}{C_{A2} - \frac{k_B}{k_A + k_B} C_T}$$

$$= \ln \frac{p_{A1} - \frac{k_B}{k_A + k_B} \pi}{p_{A2} - \frac{k_B}{k_A + k_B} \pi} \quad (5.336)$$

REFERENCES

1. Class Notes at Imperial College of Science & Technology, London, 1965.

APPENDIX I
THE GENERAL DIFFUSION EQUATION

The material balance for the conservation of mass in a small volume follows after the general material balance which states that

Rate of accumulation = *Rate of inflow*
    + *rate of generation*
    − *Rate of outflow*   (1)

Consider a small element in which mass transfer is taking place.

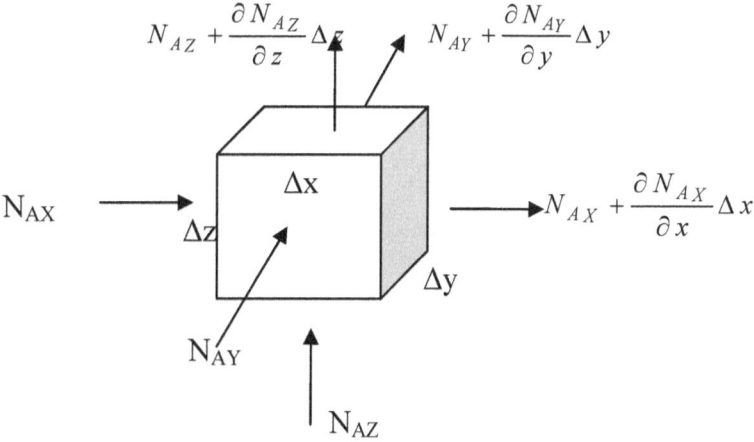

According to equation (1) and from the diagram above

$$\frac{\partial}{\partial t}(\Delta x.\Delta y.\Delta z.C_A) = N_{AX}.\Delta y.\Delta z + N_{AY}.\Delta x.\Delta z$$
$$+ N_{AZ}.\Delta x.\Delta y + R_A.\Delta x.\Delta y.\Delta z$$
$$- \left(N_{AX} + \frac{\partial N_{AX}}{\partial x}\Delta x\right)\Delta y.\Delta z$$
$$- \left(N_{AY} + \frac{\partial N_{AY}}{\partial y}\Delta y\right)\Delta x.\Delta z$$
$$- \left(N_{AZ} + \frac{\partial N_{AZ}}{\partial z}\Delta z\right)\Delta x.\Delta y \quad (2)$$

N_{AX}, N_{AY} and N_{AZ} are the molar flux of component A in the x, y and z directions, respectively. R_A is the rate of production of component A per unit volume. Simplification of equation (2) gives

$$(\Delta x.\Delta y.\Delta z.)\frac{\partial C_A}{\partial t} = R_A.\Delta x.\Delta y.\Delta z$$

$$- \frac{\partial N_{AX}}{\partial x}\Delta x.\Delta y.\Delta z$$

$$- \frac{\partial N_{AY}}{\partial y}\Delta y.\Delta x.\Delta z$$

$$- \frac{\partial N_{AZ}}{\partial z}\Delta z\Delta x.\Delta y \quad (3)$$

Further simplification yields

$$\frac{\partial C_A}{\partial t} = R_A - \frac{\partial N_{AX}}{\partial x} - \frac{\partial N_{AY}}{\partial y} - \frac{\partial N_{AZ}}{\partial z} \quad (4)$$

We know that, generally,

$$N_{AX} = U_X C_A - D_{AX}\frac{\partial C_A}{\partial x} \quad (5)$$

$$N_{AY} = U_Y C_A - D_{AY}\frac{\partial C_A}{\partial y} \quad (6)$$

$$N_{AZ} = U_Z C_A - D_{AZ}\frac{\partial C_A}{\partial z} \quad (7)$$

U_X, U_Y and U_Z and D_{AX}, D_{AY} and D_{AZ} are the bulk velocity and molar diffusivity, respectively, in the x, y and z directions. Therefore

$$\frac{\partial C_A}{\partial t} = R_A - \frac{\partial}{\partial x}\left(U_X C_A - D_{AX}\frac{\partial C_A}{\partial x}\right)$$

$$- \frac{\partial}{\partial y}\left(U_Y C_A - D_{AY}\frac{\partial C_A}{\partial y}\right) - \frac{\partial}{\partial z}\left(U_Z C_A - D_{AZ}\frac{\partial C_A}{\partial z}\right)$$

That is

APPENDIX I: THE GENERAL DIFFUSION EQUATION

$$\frac{\partial C_A}{\partial t} = R_A - U_X \frac{\partial C_A}{\partial x} + D_{AX} \frac{\partial^2 C_A}{\partial x^2}$$

$$- U_Y \frac{\partial C_A}{\partial y} + D_{AY} \frac{\partial^2 C_A}{\partial y^2} - U_Z \frac{\partial C_A}{\partial z} + D_{AZ} \frac{\partial^2 C_A}{\partial z^2}$$

This can, now, be expressed in its most familiar form as

$$\frac{\partial C_A}{\partial t} + U_X \frac{\partial C_A}{\partial x} + U_Y \frac{\partial C_A}{\partial y} + U_Z \frac{\partial C_A}{\partial z}$$

$$= D_{AX} \frac{\partial^2 C_A}{\partial x^2} + D_{AY} \frac{\partial^2 C_A}{\partial y^2} + D_{AZ} \frac{\partial^2 C_A}{\partial z^2} + R_A \quad (8)$$

There are a number of situations for which the solution of equation (8) becomes a bit simplified. These are

1. a system not in bulk flow, that is $U_X = 0$, $U_Y = 0$, $U_Z = 0$.
2. a system in steady state, that is $\frac{\partial C_A}{\partial t} = 0$
3. a system in which no chemical reaction is occurring, that is $R_A = 0$
4. a system combining one or more of items 1, 2, 3 above.

For a system not in bulk flow, $U_X = 0$, $U_Y = 0$, $U_Z = 0$.

$$\frac{\partial C_A}{\partial t} = D_{AX} \frac{\partial^2 C_A}{\partial x^2} + D_{AY} \frac{\partial^2 C_A}{\partial y^2} + D_{AZ} \frac{\partial^2 C_A}{\partial z^2} + R_A \quad (9)$$

For a system not in bulk flow and with no chemical reaction, $R_A = 0$

$$\frac{\partial C_A}{\partial t} = D_{AX} \frac{\partial^2 C_A}{\partial x^2} + D_{AY} \frac{\partial^2 C_A}{\partial y^2} + D_{AZ} \frac{\partial^2 C_A}{\partial z^2} \quad (10)$$

For a system not in bulk flow, with no chemical reaction, and unidirectional, say, in the z-direction

$$\frac{\partial C_A}{\partial t} = D_{AZ} \frac{\partial^2 C_A}{\partial z^2} \quad (11)$$

For a system not in bulk flow, with no chemical reaction,

uni-directional, say, in the z-direction and in steady state, $\dfrac{\partial C_A}{\partial t} = 0$

$$D_{AZ}\dfrac{\partial^2 C_A}{\partial z^2} = 0 \qquad (12)$$

Equations similar to equations (9), (10), (11) and (12) may be derived for mass transfer with chemical reaction.

APPENDIX II
THE ERROR FUNCTION

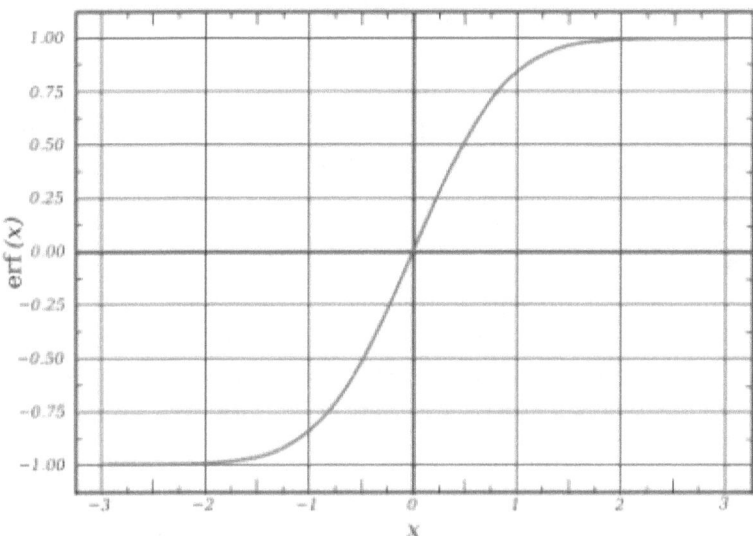

Plot Of The Error Function

In mathematics, the **error function** (also called the **Gauss error function**) is a non-elementary function which occurs in probability, statistics and partial differential equations. It is defined as:

$$erf(x) = \frac{2}{\sqrt{\pi}} \int_0^x e^{-t^2} dt$$

The **complementary error function**, denoted erfc, is defined in terms of the error function:

$$erfc(x) = 1 - erf(x) = \frac{2}{\sqrt{\pi}} \int_x^\infty e^{-t^2} dt$$

The **complex error function**, denoted $w(x)$, (also known as the Faddeeva function) is also defined in terms of the error function:

$$\omega(x) = e^{-x^2} erfc(-ix)$$

Properties

The error function is odd:

$$erf(-x) = -erf(x)$$

Also, for any complex number x one has
$$erf(x^*) = -erf(x)^*$$
where x^* is the complex conjugate of x.

The integral cannot be evaluated in closed form in terms of elementary functions, but by expanding the integrand in a Taylor series, one obtains the Taylor series for the error function as follows:

$$erf(x) = \frac{2}{\sqrt{\pi}} \sum_{n=0}^{\infty} \frac{(-1)^n x^{2n+1}}{n!(2n+1)} = \frac{2}{\sqrt{\pi}} \left(x - \frac{x^3}{3} + \frac{x^5}{10} - \frac{x^7}{42} + \frac{x^9}{216} - \cdots \right)$$

which holds for every real number x, and also throughout the complex plane. This result arises from the Taylor series expansion of e^{-x^2} which is $\sum_{n=0}^{\infty} \frac{(-1)^n x^{2n}}{n!}$ and is then integrated term by term. For iterative calculation of the above series, the following alternate formulation may be useful:

$$erf(x) = \frac{2}{\sqrt{\pi}} \sum_{n=0}^{\infty} \left(x \prod_{i=1}^{n} \frac{-(2i-1)x^2}{i(2i+1)} \right) = \frac{2}{\sqrt{\pi}} \sum_{n=0}^{\infty} \frac{x}{2n+1} \prod_{i=1}^{n} \frac{-x^2}{i}$$

because $\frac{-(2i-1)x^2}{i(2i+1)}$ expresses the multiplier to turn the ith term into the $(i+1)^{th}$ term (assuming we number the "x" as the first term). The error function at infinity is exactly 1 (see Gaussian integral). The derivative of the error function follows immediately from its definition:

$$\frac{d}{dx} erf(x) = \frac{2}{\sqrt{\pi}} e^{-x^2}.$$

An antiderivative of the error function is

$$x\, erf(x) + \frac{e^{-x^2}}{\sqrt{\pi}}.$$

The **inverse error function** has series

$$erf^{-1}(x) = \sum_{k=0}^{\infty} \frac{c_k}{2k+1} \left(\frac{\sqrt{\pi}}{2} x \right)^{2k+1},$$

where $c_0 = 1$ and

APPENDIX II: THE ERROR FUNCTION

$$c_k = \sum_{m=0}^{k-1} \frac{c_m c_{k-1-m}}{(m+1)(2m+1)} = \left\{1, 1, \frac{7}{6}, \frac{127}{90}, \ldots\right\}.$$

So we have the series expansion (note that common factors have been canceled from numerators and denominators):

$$\text{erf}^{-1}(x) = \frac{1}{2}\sqrt{\pi}\left(x + \frac{\pi}{12}x^3 + \frac{7\pi^2}{480}x^5 + \frac{127\pi^3}{40320}x^7 + \frac{4369\pi^4}{5806080}x^9 + \frac{34807\pi^5}{182476800}x^{11} + \cdots\right).$$

[1]

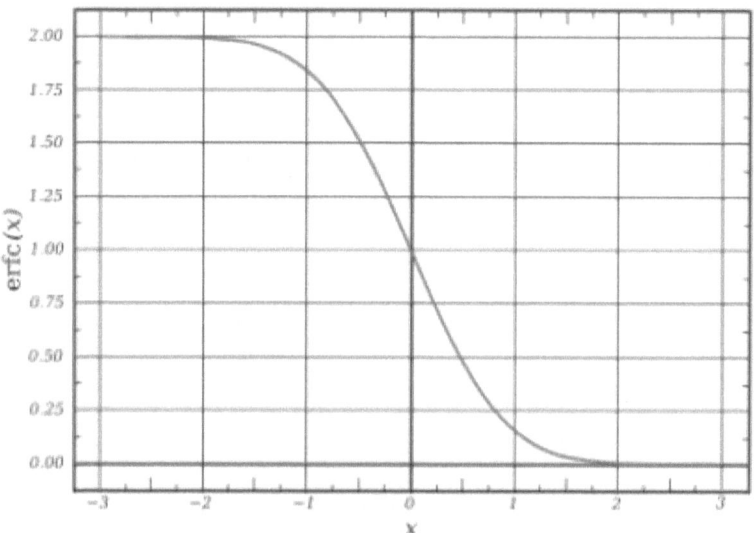

Plot Of The Complementary Error Function

Note that error function's value at plus/minus infinity is equal to plus/minus 1.

Applications

When the results of a series of measurements are described by a normal distribution with standard deviation σ and expected value 0, then $erf\left(\frac{a}{\sigma\sqrt{2}}\right)$ is the probability that the error of a single measurement lies between −a and +a, for positive a.

The error and complementary error functions occur, for example, in solutions of the heat equation when boundary conditions are given by the Heaviside step function.

Asymptotic Expansion

A useful asymptotic expansion of the complementary error function (and therefore also of the error function) for large x is

$$\text{erfc}(x) = \frac{e^{-x^2}}{x\sqrt{\pi}}\left[1+\sum_{n=1}^{\infty}(-1)^n\frac{1\cdot 3\cdot 5\cdots(2n-1)}{(2x^2)^n}\right] = \frac{e^{-x^2}}{x\sqrt{\pi}}\sum_{n=0}^{\infty}(-1)^n\frac{(2n)!}{n!(2x)^{2n}}.$$

This series diverges for every finite x. However, in practice only the first few terms of this expansion are needed to obtain a good approximation of erfc(x), whereas the Taylor series given above converges very slowly.

Another approximation is given by

$$\text{erf}^2(x) \approx 1 - \exp\left(-x^2\frac{4/\pi + ax^2}{1+ax^2}\right)$$

where

$$a = -\frac{8(\pi-3)}{3\pi(\pi-4)}.$$

References

1. Milton Abramowitz and Irene A. Stegun, eds. *Handbook of Mathematical Functions with Formulas, Graphs, and Mathematical Tables.* New York: Dover, 1972. *(See Chapter 7)*
2. MathWorld - Erf

Retrieved from "http://en.wikipedia.org/wiki/Error_function"

This page was last modified on 16 June 2008, at 14:26.

APPENDIX III
THE LAPLACE TRANSFORM

Formal Definition

The Laplace transform of a function $f(t)$, defined for all real numbers $t \geq 0$, is the function $F(s)$, defined by:

$$F(s) = \mathcal{L}\{f(t)\} = \int_{0^-}^{\infty} e^{-st} f(t)\, dt.$$

The lower limit of 0^- is short notation to mean

$$\lim_{\varepsilon \to 0+} \int_{-\varepsilon}^{\infty}$$

and assures the inclusion of the entire Dirac delta function $\delta(t)$ at 0 if there is such an impulse in $f(t)$ at 0.

The parameter s is in general complex:

$$s = \sigma + i\omega$$

This integral transform has a number of properties that make it useful for analyzing linear dynamic systems. The most significant advantage is that differentiation and integration become multiplication and division, respectively, by s. (This is similar to the way that logarithms change an operation of multiplication of numbers to addition of their logarithms.) This changes integral equations and differential equations to polynomial equations, which are much easier to solve. Once solved, use of the inverse Laplace transform reverts back to the time domain.

Bilateral Laplace Transform

When one says "the Laplace transform" without qualification, the unilateral or one-sided transform is normally intended. The Laplace transform can be alternatively defined as the *bilateral*

Laplace transform or two-sided Laplace transform by extending the limits of integration to be the entire real axis. If that is done the common unilateral transform simply becomes a special case of the bilateral transform where the definition of the function being transformed is multiplied by the Heaviside step function.

The bilateral Laplace transform is defined as follows:

$$F(s) = \mathcal{L}\{f(t)\} = \int_{-\infty}^{+\infty} e^{-st} f(t)\, dt.$$

Inverse Laplace Transform

The inverse Laplace transform is given by the following complex integral, which is known by various names (the **Bromwich integral**, the **Fourier-Mellin integral**, and **Mellin's inverse formula**):

$$f(t) = \mathcal{L}^{-1}\{F(s)\} = \frac{1}{2\pi i} \int_{\gamma-i\cdot\infty}^{\gamma+i\cdot\infty} e^{st} F(s)\, ds,$$

where γ is a real number so that the contour path of integration is in the *region of convergence* of $F(s)$ normally requiring $\gamma > \mathrm{Re}(s_p)$ for every singularity s_p of $F(s)$ and $i^2 = -1$. If all singularities are in the left half-plane, that is $\mathrm{Re}(s_p) < 0$ for every s_p, then γ can be set to zero and the above inverse integral formula becomes identical to the inverse Fourier transform.

An alternative formula for the inverse Laplace transform is given by Post's inversion formula.

Region Of Convergence

The Laplace transform $F(s)$ typically exists for all complex numbers such that $\mathrm{Re}\{s\} > a$, where a is a real constant which depends on the growth behavior of $f(t)$, whereas the two-sided transform is defined in a range $a < \mathrm{Re}\{s\} < b$. The subset of values of s for which the Laplace transform exists is called the

APPENDIX III: THE LAPLACE TRANSFORM

region of convergence (ROC) or the *domain of convergence*. In the two-sided case, it is sometimes called the *strip of convergence*.

The integral defining the Laplace transform of a function may fail to exist for various reasons. For example, when the function has infinite discontinuities in the interval of integration, or when it increases so rapidly that exp(-pt) cannot damp it sufficiently for convergence on the interval to take place. There are no specific conditions that one can check a function against to know in all cases if its Laplace transform can be taken, other than to say the defining integral converges. It is however easy to give theorems on cases where it may or may not be taken.

Properties And Theorems

Given the functions $f(t)$ and $g(t)$, and their respective Laplace transforms $F(s)$ and $G(s)$:

$$f(t) = \mathcal{L}^{-1}\{F(s)\}$$
$$g(t) = \mathcal{L}^{-1}\{G(s)\}$$

the following table is a list of properties of unilateral Laplace transform:

Properties Of The Unilateral Laplace Transform

	Time domain	Frequency domain	Comment
Linearity	$a f(t) + b g(t)$	$a F(s) + b G(s)$	Can be proved using basic rules of integration.

Frequency differentiation	$t\,f(t)$	$-F'(s)$	
Frequency differentiation	$t^n f(t)$	$(-1)^n F^{(n)}(s)$	More general form
Differentiation	$f'(t)$	$sF(s) - f(0^-)$	Write the exact integral form of the given function, and add another integral to complement the former to deduce the sum to indefinite integration of a differential. Next few steps are simple.
Second Differentiation	$f''(t)$	$s^2 F(s)$ $-s\,f(0^-)$ $-f'(0)$	Apply the Differentiation property to $f(t)$.
General Differentiation	$f^{(n)}(t)$	$s^n F(s)$ $-s^{n-1} f(0^-)$ $-\cdots-$	Follow the process briefed for the Second Differentiation.
Frequency integration	$\dfrac{f(t)}{t}$	$\int_s^\infty F(\sigma)\,d\sigma$	

APPENDIX III: THE LAPLACE TRANSFORM

Integration	$\int f(\tau)d\tau$ $= u(t) * f(t)$	$\dfrac{1}{s}F(s)$	$u(t)$ is the Heaviside step function. Note $u(t) * f(t)$ is the convolution of $u(t)$ and $f(t)$, not multiplication.
Scaling	$f(at)$	$\dfrac{1}{\|a\|}F(as)$	
Frequency shifting	$e^{at}f(t)$	$F(s-a)$	
Time shifting	$f(t-a)u(t-a)$	$e^{-as}F(s)$	$u(t)$ is the Heaviside step function
Convolution	$(f*g)(t)$	$F(s).G(s)$	
Periodic Function	$f(t)$	$\dfrac{1}{1-e^{-Ts}}\int_0^T e^{-st}$	$f(t)$ is a periodic function of period T so that $f(t) = f(t+T)$,

- **Initial value theorem**:

$$f(0^+) = \lim_{s \to \infty} sF(s)$$

- **Final value theorem:**

$$f(\infty) = \lim_{s \to 0} sF(s)$$, if all poles of $sF(s)$ are in the left-hand plane.

The final value theorem is useful because it gives the long-term behaviour without having to perform partial fraction decompositions or other difficult algebra. If a function's poles are in the right hand plane (e.g. e^t or sin(*t*)) the behaviour of this formula is undefined.

Proof Of The Laplace Transform Of A Function's Derivative

It is often convenient to use the differentiation property of the Laplace transform to find the transform of a function's derivative. This can be derived from the basic expression for a Laplace transform as follows:

$$\mathcal{L}\{f(t)\} = \int_{0^-}^{+\infty} e^{-st} f(t)\, dt$$

$$= \left[\frac{f(t)e^{-st}}{-s}\right]_{0^-}^{+\infty} - \int_{0^-}^{+\infty} \frac{e^{-st}}{-s} f'(t)\, dt \quad \text{(by parts)}$$

$$= \left[-\frac{f(0)}{-s}\right] + \frac{1}{s}\mathcal{L}\{f'(t)\},$$

yielding

$$\mathcal{L}\left\{\frac{df}{dt}\right\} = s \cdot \mathcal{L}\{f(t)\} - f(0),$$

and in the bilateral case, we have

$$\mathcal{L}\left\{\frac{df}{dt}\right\} = s \int_{-\infty}^{+\infty} e^{-st} f(t)\, dt = s \cdot \mathcal{L}\{f(t)\}.$$

APPENDIX III: THE LAPLACE TRANSFORM

Table Of Selected Laplace Transforms

The following table provides Laplace transforms for many common functions of a single variable. For definitions and explanations, see the *Explanatory Notes* at the end of the table.

Because the Laplace transform is a linear operator:

- The Laplace transform of a sum is the sum of Laplace transforms of each term.

$$\mathcal{L}\{f(t) + g(t)\} = \mathcal{L}\{f(t)\} + \mathcal{L}\{g(t)\}$$

- The Laplace transform of a multiple of a function, is that multiple times the Laplace transformation of that function.

$$\mathcal{L}\{af(t)\} = a\mathcal{L}\{f(t)\}$$

The unilateral Laplace transform is only valid when t is non-negative, which is why all of the time domain functions in the table below are multiples of the Heaviside step function, u(*t*).

ID	Function	Time domain $x(t) = L^{-1}\{X(s)\}$	Laplace s-domain $X(s) = L\{x(t)\}$	Region of convergence for causal systems
1	ideal delay	$\delta(t - \tau)$	$e^{-\tau s}$	
1a	unit impulse	$\delta(t)$	1	All s

2	delayed nth power with frequency shift	$\dfrac{(t-\tau)^n}{n!} e^{-\alpha(t-\tau)} \cdot u(t-\tau)$	$\dfrac{e^{-\tau s}}{(s+\alpha)^{n+1}}$	$\mathrm{Re}\{s\}>0$
2a	nth power (for integer n)	$\dfrac{t^n}{n!} \cdot u(t)$	$\dfrac{1}{s^{n+1}}$	$\mathrm{Re}\{s\}>0$
2a.1	qth power (for real q)	$\dfrac{t^q}{\Gamma(q+1)} \cdot u(t)$	$\dfrac{1}{s^{q+1}}$	$\mathrm{Re}\{s\}>0$
2a.2	unit step	$u(t)$	$\dfrac{1}{s}$	$\mathrm{Re}\{s\}>0$
2b	delayed unit step	$u(t-\tau)$	$\dfrac{e^{-\tau s}}{s}$	$\mathrm{Re}\{s\}>0$
2c	ramp	$t \cdot u(t)$	$\dfrac{1}{s^2}$	$\mathrm{Re}\{s\}>0$
2d	nth power with frequency shift	$\dfrac{t^n}{n!} e^{-\alpha t} \cdot u(t)$	$\dfrac{1}{(s+\alpha)^{n+1}}$	$\mathrm{Re}\{s\}>-\alpha$
2d.1	exponential decay	$e^{-\alpha t} \cdot u(t)$	$\dfrac{1}{s+\alpha}$	$\mathrm{Re}\{s\}>-\alpha$

APPENDIX III: THE LAPLACE TRANSFORM

3	Exponential approach	$(1 - e^{-\alpha t}) \cdot u(t)$	$\dfrac{\alpha}{s(s+\alpha)}$	$\text{Re}\{s\} > 0$
4	sine	$\sin(\omega t) \cdot u(t)$	$\dfrac{\omega}{s^2 + \omega^2}$	$\text{Re}\{s\} > 0$
5	cosine	$\cos(\omega t) \cdot u(t)$	$\dfrac{s}{s^2 + \omega^2}$	$\text{Re}\{s\} > 0$
6	hyperbolic sine	$\sinh(\alpha t) \cdot u(t)$	$\dfrac{\alpha}{s^2 - \alpha^2}$	$\text{Re}\{s\} > |\alpha|$
7	hyperbolic cosine	$\cosh(\alpha t) \cdot u(t)$	$\dfrac{s}{s^2 - \alpha^2}$	$\text{Re}\{s\} > |\alpha|$
8	Exponentially-decaying sine wave	$e^{\alpha t} \sin(\omega t) \cdot u(t)$	$\dfrac{\omega}{(s-\alpha)^2 + \omega^2}$	$\text{Re}\{s\} > \alpha$
9	Exponentially-decaying cosine wave	$e^{\alpha t} \cos(\omega t) \cdot u(t)$	$\dfrac{s-\alpha}{(s-\alpha)^2 + \omega^2}$	$\text{Re}\{s\} > \alpha$

10	nth root	$\sqrt[n]{t} \cdot u(t)$	$s^{-\frac{n+1}{n}} \cdot \Gamma\left(1+\frac{1}{n}\right)$	$\text{Re}\{s\}>0$
11	natural logarithm	$\ln\left(\dfrac{t}{t_0}\right) \cdot u(t)$	$-\dfrac{t_0}{s}\left[\ln(t_0 s) + \gamma\right]$	$\text{Re}\{s\}>0$
12	Bessel function of the first kind, of order n	$J_n(\omega t) \cdot u(t)$	$\dfrac{\omega^n\left(s + \sqrt{s^2 + \omega^2}\right)^{-n}}{\sqrt{s^2 + \omega^2}}$	$\text{Re}\{s\}>0$ $(n>-1)$
13	Modified Bessel function of the first kind, of order n	$I_n(\omega t) \cdot u(t)$	$\dfrac{\omega^n\left(s + \sqrt{s^2 - \omega^2}\right)^{-n}}{\sqrt{s^2 - \omega^2}}$	$\text{Re}\{s\} >$
14	Bessel function of the second kind, of order 0	$Y_0(\alpha t) \cdot u(t)$	$-\dfrac{2\sinh^{-1}\left(\dfrac{s}{\alpha}\right)}{\pi\sqrt{s^2 + \alpha^2}}$	$\text{Re}\{s\}>0$
15	Modified Bessel function of the second kind, of order 0	$K_0(\alpha t) \cdot u(t)$		

APPENDIX III: THE LAPLACE TRANSFORM

| 16 | Error function | $\mathrm{erf}(t) \cdot u(t)$ | $\dfrac{e^{\frac{s^2}{4}} \left(1 - \mathrm{erf}\left(\dfrac{s}{2}\right)\right)}{s}$ | $\mathrm{Re}\{s\} > 0$ |

Explanatory notes:

- $u(t)$ represents the Heaviside step function.
- $\delta(t)$ represents the Dirac delta function.
- $\Gamma(z)$ represents the Gamma function.
- γ is the Euler-Mascheroni constant.

- t, a real number, typically represents *time*, although it can represent *any* independent dimension.
- s is the complex angular frequency, and $\mathrm{Re}\{s\}$ is its real part.
- α, β, τ, and ω are real numbers.
- n, is an integer.

- A causal system is a system where the impulse response $h(t)$ is zero for all time t prior to $t = 0$. In general, the ROC for causal systems is not the same as the ROC for anticausal systems.

Retrieved from "http://en.wikipedia.org/wiki/Laplace_transform"

APPENDIX IV
SURFACE TENSION

Inorganic. $\gamma = a - bt$, mN m^{-1}

Substance	t_m °C	γ mN m^{-1}	Equation constants		
			a	b	Range °C
Lead chloride . .	520	135.3	199.8	0.124	510–580
Potassium chloride .	780	100.3	160.4	0.070	770–970
Potassium nitrate .	340	111.0	136.5	0.0750	340–500
Sodium chloride .	810	113.3	171.5	0.0719	805–970
Sodium fluoride .	1000	185.2	267.2	0.082	1000–1080
Sodium nitrate. . .	320	119.2	138.8	0.0613	320–600
Sodium sulphate . .	900	194.5	239.6	0.0501	900–1080
Oxygen	−184	13.40	−33.72	0.2561	−202 to −184
Nitrogen	−183	5.99	−35.48	0.2266	−195 to −183

More data on molten salt in Janz (1969).

Organic $\gamma = a - bt$, mN m^{-1}

Substance	Temp. °C	γ mN m^{-1}	Equation constants		
			a	b	Range °C
Acetic acid . . .	20	27.59	29.58	0.099 4	20–90

Acetone	20	23.46	26.26	0.112	20–50
Aniline	20	42.67	44.83	0.108 5	15–90
Benzene	20	28.88	31.50	0.128 7	10–80
n-Butanol. . . .	20	25.39	27.18	0.089 83	10–100
Carbon disulphide. .	20	32.32	35.29	0.148 4	10–50
Carbon tetrachloride	20	27.04	29.49	0.122 4	15–105
Chloroform . . .	20	27.32	29.91	0.129 5	15–75
Di-ethyl ether . . .	20	17.10	18.92	0.090 8	15–30
Ethanol	20	22.39	24.05	0.083 2	10–70
Ethyl acetate . . .	20	23.97	26.29	0.116 1	10–100
Glycerol	20	63.4	65.17	0.088 45	20–150
n-Hexane	20	18.40	20.44	0.102 2	10–60
n-Octane	20	21.62	23.52	0.095 09	10–120
Methanol	20	22.50	24.00	0.077 3	10–60
Methyl acetate . .	20	25.37	27.95	0.128 9	10–60
Phenol	40	39.27	43.54	0.106 8	40–140
n-Propanol . . .	20	23.71	25.26	0.077 7	10–90
Toluene	20	28.52	30.90	0.188 9	10–100

For more data see Jasper (1972).

Surface Tension Of Water Against Air

Temp/ °C	0	10	15	20	25	30	40	50	60	70	80	100
$\gamma/(mN\ m^{-1})$	75.7	74.2	73.5	72.75	72.0	71.2	69.6	67.9	66.2	64.4	62.6	58.8

APPENDIX IV: SURFACE TENSION

Surface Tensions Of Aqueous Salt Solutions.

Usually greater than that of water. Approximately $\gamma = \gamma H_2O + M \cdot \Delta\gamma$, where M is concentration. Values below are for $M = 1$ mole dm^{-3} at 20 °C.

Salt	$\Delta\gamma$/(mN m^{-1})
KCl	1.4
NaCl	1.64
Na$_2$CO$_3$	2.7
NaNO$_3$	1.2
Na$_2$SO$_4$	2.7

Interfacial Tensions Of Liquids At 20 °C

Liquids	γ/(mN m^{-1})	Liquids	γ/(mN m^{-1})
Water against:		Mercury against:	
Benzene	35	Acetone	390
Carbon tetrachloride	45	Benzene	357
Chloroform	28	Chloroform	357
Di-ethyl ether	10	Di-ethyl ether	379
Heptylic acid	7	n-Heptane	379
n-Heptane	51	Oleic acid	322
n-Octane	51		
Olive oil	20		
Paraffin oil	48		

References

1. Adam, *The Physics and Chemistry of Surfaces.*

2. Eötvös (1886).

3. G. J. Janz (1969) *NSRDS-NBS 28*, report, Natl. Bur. Standards, Washington.

4. J. J. Jasper (1972) *J. Phys. Chem. Ref. Data*, **I**(4), 841–1010.

B.J.Keene

Surface Tension Values Of Some Common Test Liquids For Surface Energy Analysis

Name	CAS Ref.-No.	Surface tension @ 20 °C in mN/m	Temperature coefficient in mN/(m K)
1,2-Dichloro ethane	107-06-2	33.30	-0.1428
1,2,3-Tribromo propane	96-11-7	45.40	-0.1267
1,3,5-Trimethylbenzene (Mesitylene)	108-67-8	28.80	-0.0897
1,4-Dioxane	123-91-1	33.00	-0.1391
1,5-Pentanediol	111-29-5	43.30	-0.1161
1-Chlorobutane	109-69-3	23.10	-0.1117
1-Decanol	112-30-1	28.50	-0.0732
1-nitro propane	108-03-2	29.40	-0.1023
1-Octanol	111-87-5	27.60	-0.0795
Acetone (2-Propanone)	67-64-1	25.20	-0.1120
Aniline 22°C (AN)	62-53-3	43.40	-0.1085
2-Aminoethanol	141-43-5	48.89	-0.1115
Anthranilic acid ethylester 22°C	87-25-2	39.30	-0.0935

APPENDIX IV: SURFACE TENSION

Anthranilic acid methylester 25 °C	134-20-3	43.71	-0.1152
Benzene	71-43-2	28.88	-0.1291
Benzylalcohol	100-51-6	39.00	-0.0920
Benzylbenzoate (BNBZ)	120-51-4	45.95	-0.1066
Bromobenzene	108-86-1	36.50	-0.1160
Bromoform	75-25-2	41.50	-0.1308
Butyronitrile	109-74-0	28.10	-0.1037
Carbon disulfid	75-15-0	32.30	-0.1484
Quinoline	91-22-5	43.12	-0.1063
Chloro benzene	108-90-7	33.60	-0.1191
Chloroform	67-66-3	27.50	-0.1295
Cyclohexane	110-82-7	24.95	-0.1211
Cyclohexanol 25 °C	108-93-0	34.40	-0.0966
Cyclopentanol	96-41-3	32.70	-0.1011
p-Cymene	99-87-6	28.10	-0.0941
Decalin	493-01-6	31.50	-0.1031
Dichloromethane	75-09-2	26.50	-0.1284
Diiodomethane (DI)	75-11-6	50.80	-0.1376
1,3-Diiodopropane 23 °C	627-31-6	46.51	-0.1195
Diethylene glycol (DEG)	111-46-6	44.80	-0.0841
Dipropylene glycol	25265-71-8	33.90	-0.1070
Dipropylene glycol monomethylether	34590-94-8	28.41	-0.1088
Dodecyl benzene	123-01-3	30.70	
Ethanol	64-17-5	22.10	-0.0832
Ethylbenzene	100-41-4	29.20	-0.1094
Ethylbromide	74-96-4	24.20	-0.1159
Ethylene glycol (EG)	107-21-1	47.70	-0.0890
Formamide (FA)	75-12-7	58.20	-0.0842
Fumaric acid diethylester 22°C	623-91-6	31.40	-0.1039
Furfural (2-Furaldehyde)	98-01-1	41.90	-0.1225
Glycerol (GLY)	56-81-5	64.00	-0.0598

Ethylene glycol monoethyl ether (Ethyl Cellosolve)	110-80-5	28.60	-0.0918
Hexachlorobutadiene	87-68-3	36.00	-0.0994
Iodobenzene	591-50-4	39.70	-0.1123
Isoamylchloride	107-84-6	23.50	-0.1078
Isobutylchloride	513-36-0	21.90	-0.1144
Isopropanol	67-63-0	23.00	-0.0789
Isopropylbenzene	98-82-8	28.21	-0.1054
Isovaleronitrile	625-28-5	26.00	-0.0827
m-Nitrotoluene	99-08-1	41.40	-0.1140
Mercury	7439-97-6	425.41	-0.2049
Methanol	67-56-1	22.70	-0.0773
Methyl ethyl ketone (MEK)	78-93-3	24.60	-0.1199
Methyl naphthalene	90-12-0	38.60	-0.1118
N,N-dimethyl acetamide (DMA)	127-19-5	36.70	-0.1395
N,N-dimethyl formamide (DMF)	68-12-2	37.10	-0.1400
N-methyl-2-pyrrolidone	872-50-4	40.79	-0.1156
n-Decane (DEC)	124-18-5	23.83	-0.0920
n-Dodecane (DDEC)	112-40-3	25.35	-0.0884
n-Heptane	142-82-5	20.14	-0.0980
n-Hexadecane (HDEC)	544-76-3	27.47	-0.0854
n-Hexane (HEX)	110-54-3	18.43	-0.1022
n-Octane (OCT)	111-65-9	21.62	-0.0951
n-Tetradecane (TDEC)	629-59-4	26.56	-0.0869
n-Undecane	1120-21-4	24.66	-0.0901
n-Butylbenzene	104-51-8	29.23	-0.1082
n-Propylbenzene	103-65-1	28.99	-0.1071
Nitroethane	79-24-3	31.90	-0.1255
Nitrobenzene	98-95-3	43.90	-0.1177
Nitromethane	75-52-5	36.80	-0.1678
o-Nitrotoluene	88-72-2	41.50	-0.1229

APPENDIX IV: SURFACE TENSION

Substance	CAS	Value	Coeff.
Perfluoroheptane	335-57-9	12.85	-0.0972
Perfluorohexane	355-42-0	11.91	-0.0935
Perfluorooctane	307-34-6	14.00	-0.0902
Phenylisothiocyanate	103-72-0	41.50	-0.1172
Phthalic acid diethylester 22°C	84-66-2	37.00	-0.1018
Polyethylen glycol 200 (PEG)	25322-68-3	43.50	-0.1170
Polydimethyl siloxane (Baysilone M5)	9016-00-6	19.00	-0.0365
Propanol 25 °C	71-23-8	23.70	-0.0777
Pyridine	110-86-1	38.00	-0.1372
3-Pyridylcarbinol 23°C (PYC)	100-55-0	47.68	-0.1259
Pyrrol (PY)	109-97-7	36.60	-0.1100
sym-Tetrabromoethane	79-27-6	49.70	-0.1528
tert-Butylchloride	507-20-0	19.60	-0.1072
sym-Tetrachloromethane	56-23-5	26.95	-0.1224
Tetrahydrofuran (THF)	109-99-9	26.40	-0.1277
Thiodiglycol (2,2'-Thiobisethanol) (TDG)	111-48-8	54.00	-0.0830
Toluene	108-88-3	28.40	-0.1189
Tricresylphosphate (TCP)	1330-78-5	40.90	-0.0887
Water (WA)	7732-18-5	72.80	-0.1514
o-Xylene	95-47-6	30.10	-0.1101
m-Xylene	108-38-3	28.90	-0.1104
a-Bromonaphthalene (BN)	90-11-9	44.40	-0.0979
a-Chloronaphthalene	90-13-1	41.80	-0.1064

Source: http://www.dataphysics.de (Europe, Asia and worldwide) or http://www.fdsc.com (USA).

Last updated: 24 Nov 2006 - www.surface-tension.de

APPENDIX V
VISCOSITY TABLES

The **dynamic viscosity,** μ of a (Newtonian) fluid is given by $\mu = \tau \div dv/dr$; τ = shearing stress between two planes parallel with the direction of flow, dv/dr = velocity gradient at right angles to the direction of flow.
The dimensions of dynamic viscosity are $ML^{-1}T^{-1}$, and the SI unit is Pa s.

Kinematic viscosity, v, is the ratio of the dynamic viscosity to the density, ρ. The dimensions of kinematic viscosity are L^2T^{-1} and the SI unit is $m^2 s^{-1}$.

Fluidity, ϕ is the reciprocal of the dynamic viscosity, μ. The dynamic viscosity of liquids decreases with the temperature approximately according to the equation
$$\log \eta = A + B/T,$$
where A and B are characteristic constants and T is the absolute temperature.

Values of A and B for a large number of liquids are given by Barrer (1943).

(i) Viscosity of water. Data from Kestin, Sokolov and Wakeham (1978).

Temp. °C	μ mPa s	Temp. °C	μ mPa s	Temp. °C	μ mPa s	Temp. °C	μ mPa s
0	1.792	20	1.0020	50	0.5471	90	0.3150
5	1.519	25	0.8902	60	0.4670	100	0.2821
10	1.307	30	0.7973	70	0.4046	125	0.2217
15	1.138	40	0.6526	80	0.3551	150	0.1818

(Ii) Viscosities Of Various Liquids. μ In Mpa S

Liquid	−100 °C	−50 °C	0 °C	25 °C	30 °C	50 °C	75 °C	100 °C
Acetic acid	—	—	—	1.116	1.037	0.792	0.591	0.457
Acetone	—	—	0.402	0.310	0.295	0.247	0.200	0.165
Aniline	—	—	9.450	3.822	3.298	1.982	1.201	0.808
Benzene	—	—	—	0.603	0.562	0.436	0.332	0.263
Bromo Benzene	—	—	1.592	1.065	0.995	0.780	0.605	0.488
n-butane	0.747	0.339	0.197	0.157	0.150	—	—	—
Carbon disulphide	2.132	0.796	0.445	0.357	0.343	—	—	—
Carbon dioxide	—	0.227	0.098	0.057	—	—	—	—
Carbon tetrachloride	—	—	1.341	0.912	0.851	0.662	0.503	0.395
Chloroform	—	1.514	0.709	0.536	0.510	0.422	0.341	0.281
Di-ethyl ether	1.545	0.544	0.288	0.224	0.214	0.179	0.146	0.119
Ethanol	98.96	8.318	1.873	1.084	0.983	0.684	0.459	0.323
Ethyl acetate	—	1.284	0.575	0.428	0.406	0.332	0.264	0.215
Ethyl formate	—	1.060	0.504	0.381	0.362	0.299	0.240	0.196
n-Hexane	—	0.782	0.387	0.296	0.282	0.235	0.190	0.155
n-Hexadecane	—	—	—	3.044	2.729	1.852	1.245	0.896
Mercury	—	—	1.616	1.528	1.497	1.401	1.322	1.255
Methane	0.0357	—	—	—	—	—	—	—
Methanol	—	2.258	0.797	0.543	0.507	0.392	0.294	0.227
Nitrobenzene	—	—	—	1.842	1.688	1.244	0.915	0.710

APPENDIX V: VISCOSITY TABLES

n-Octane	—	1.837	0.719	0.516	0.486	0.390	0.306	0.247
Oil, castor	—	—	—	700	451	125	42.0	16.9
Oil, olive	—	—	—	67.0	54.0	25.8	9.4	7.0
n-Pentane	1.300	0.498	0.273	0.214	0.205	0.173	0.140	0.115
n-Propane	0.421	0.215	0.127	0.099	0.094	—	—	—
Sulphuric acid	—	—	—	23.8	20.1	11.7	6.6	4.1
Toluene	—	2.124	0.768	0.551	0.520	0.420	0.334	0.272

For more data see: ESDU (1966–83), Landolt-Börstein (1969).

(Iii) Viscosity Of Aqueous Glycerol Solutions.

Data from Segur and Oberstar (1951), corrected to value for water at 20 °C of 1.002 mPa s.

Density 20° kg/l	% weight glycerol	μ /Pa s		
		20°C	30°C	40°C
1.2611	100	1.408	0.610	0.283
1.2588	99	1.146	0.498	0.234
1.2562	98	0.936	0.408	0.195
1.2534	97	0.763	0.339	0.165
1.2508	96	0.622	0.280	0.142
1.2482	95	0.521	0.236	0.121
1.2085	80	0.059 9	0.033 8	0.020 7
1.1254	50	0.005 98	0.004 20	0.003 09
1.0459	20	0.001 75	0.001 35	0.001 07
1.0215	10	0.001 31	0.001 03	0.000 823

(Iv) Viscosity Of Aqueous Sucrose Solutions.

Data from Bingham and Jackson (1918), corrected to a more recent value for the viscosity of water.

Relative density 20°/4 °C	% weight sucrose	μ /Pa s		
		15 °C	20 °C	25 °C
1.3790	75	4.039	2.328	1.405
1.3472	70	0.746 9	0.481 6	0.321 6
1.3163	65	0.211 3	0.147 2	0.105 4
1.2865	60	0.079 49	0.058 49	0.040 03
1.2296	50	0.019 53	0.015 43	0.012 40
1.1764	40	0.007 463	0.006 167	0.005 164
1.1270	30	0.003 757	0.003 187	0.002 735

(V) Relative Viscosities Of Some Aqueous Solutions At 1 N Concentration.

For a complete list see International Critical Tables (1928) and Stokes and Mills (1965). (The latter covers 1929–63.)

Substance	Temp. /°C	Relative viscosity	Substance	Temp. /°C	Relative viscosity
Ammonia	25	1.02	Potassium chloride	17.6	0.98
Ammonium chloride	17.6	0.98	Potassium iodide	17.6	0.91
Calcium chloride	20	1.31	Sodium hydroxide	25	1.24
Hydrochloric	15	1.07	Sulphuric	25	1.09

acid			acid		

(Vi) Viscosity Of Liquid Metals And Molten Salts. μ In Mpa S

Liquid	100 °C	400 °C	600 °C	700 °C	800 °C	1100 °C	1200 °C
Aluminium	—	—	—	2.96	2.66	—	—
Gold	—	—	—	—	—	5.13	4.64
Lead	—	2.32	1.55	1.37	1.24	—	—
Potassium	0.458	0.224	0.172	0.155	0.141	—	—
Sodium	0.680	0.286	0.215	0.192	0.174	—	—
Tin	—	1.33	1.04	0.95	0.89	0.78	0.77
Potassium chloride	—	—	—	—	1.096	–	–
Potassium nitrate	—	2.09	1.07	—	—	—	—
Sodium chloride	—	—	—	—	1.500[†]	—	—
Sodium nitrate	—	1.91	–	—	—	—	—

[†]816°C.

For more data on molten salts see G. J. Janz (1968). For liquid metal see Smithell (1983, p. 14.2).

Viscosities Of Glasses And Minerals. $\log_{10}(\mu/\text{Pa S})$

Material	900 °C	1000 °C	1100 °C	1200 °C	1300 °C	1400 °C	1600 °C	1800 °C	2000 °C
Plate glass	4.00	3.03	2.41	1.87	1.46	1.07	—	—	—
Medium flint glass	3.9	2.8	1.9	1.4	0.9	0.7	—	—	—
Silica	—	—	14.6	12.7	11.8	9.7	8.2	4.7	3.4
Olivine	—	—	—	2.5	1.5	1.2	—	—	—

Diorite	—	—	—	3.1	2.3	1.8	—	—	—
Diopside	—	—	—	—	0.52	0.43	—	—	—

Viscosities Of Liquids At High Pressures

(I) Relative Viscosity Of Water

Pressure/MPa	2.2 °C	10 °C	20 °C	30 °C	50 °C	75 °C	100 °C
0.1	1.000	1.000	1.000	1.000	1.000	1.000	1.000
49	0.946	0.969	0.990	0.998	1.021	1.029	1.043
98	0.926	0.957	0.990	1.008	1.046	1.063	1.085
196	0.940	0.982	1.023	1.053	1.104	1.137	1.170
294	0.993	1.037	1.081	1.116	1.174	1.217	1.256
392	1.072	1.185	1.163	1.195	1.253	1.302	1.349
588	1.296	1.330	1.367	1.386	1.439	1.492	1.546
784	—	—	1.629	1.642	1.664	1.708	1.757
981	—	—	—	1.950	1.936	1.948	1.986

(Ii) Relative Viscosities Of Various Liquids. Ratio = μ_p/μ_0 At Same Temperature.

Liquid	Temp./°C	Pressure/MPa					
		98	100	392	400	784	1177
Acetone	30	1.68	—	4.03	—	9.70	—
	75	1.65	—	3.55	—	7.36	13.7
Benzene	30	2.22	—	—	—	—	—
	50	—	2.06	—	—	—	—
	75	2.07	2.00	—	—	—	—
	100	—	1.98	—	7.40	—	—
Carbon disulphide	30	1.45	—	3.23	—	6.92	15.5
	75	1.50	—	3.14	—	6.25	11.8

APPENDIX V: VISCOSITY TABLES

Di-ethyl ether	30	2.11	—	6.20	—	18.2	46.8
	75	1.87	—	5.28	—	12.8	27.1
Ethanol	30	1.59	—	4.14	—	10.5	24.5
	75	1.64	—	4.28	—	9.48	18.3
n-Hexane	25	—	2.10	—	—	—	—
	30	2.15	—	8.20	—	32.7	—
	75	2.33	2.21	7.91	7.41	24.8	69.7
	100	–	2.26	—	7.22	—	—
n-Hexadecane	50	—	2.89	—	—	—	—
	75	—	2.69	—	—	—	—
	100	—	2.61	—	15.0	—	—
Methanol	30	1.47	—	2.96	—	5.62	9.95
	75	1.46	2.74	—	—	4.77	7.69
n-Octane	25	–	2.31	—	12.9	—	—
	30	2.12	—	12.3	—	—	–
	75	2.20	2.24	8.97	9.19	35.7	—
	100	—	2.25	—	8.56	—	—
n-Pentane	30	2.07	—	7.03	—	22.9	70.2
	75	2.25	—	7.33	—	20.3	48.1
Toluene	30	1.87	7.89	—	—	50.0	—
	75	1.86	—	6.33	—	24.6	109

Data from Bridgman (1926), except at 100 and 400 MPa which comes from K. J. Young, PhD Thesis, University of Glasgow, Nov. 1980, and Dymond, Robertson and Isdale (1981).

Viscosities Of Gases And Vapours

(i) Viscosities At Normal Pressure. Units: μPa s.

Gas	0 °C	20 °C	50 °C	100 °C	200 °C	300 °C	400 °C	500 °C	600 °C
Air	17.3	18.2	19.6	22.0	26.1	29.8	33.2	36.4	39.4
Ammonia	9.2	9.9	11.0	13.0	16.8	20.6	24.4	28.2	31.9
Argon	21.0	22.3	24.2	27.3	32.8	37.7	42.2	46.4	50.4
Benzene	7.0	7.5	8.1	9.4	12.0	—	—	—	—

Gas									
Carbon dioxide	13.7	14.7	16.1	18.5	23.0	27.1	30.8	34.2	37.4
Carbon monoxide	16.6	17.4	18.8	21.0	25.2	29.0	32.5	35.6	38.6
Chlorine	12.3	13.2	14.5	16.9	21.0	25.0	—	—	—
Chloroform	9.4	10.1	11.1	12.8	16.2	19.5	—	—	—
Ethylene	9.7	10.3	11.2	12.8	15.4	17.9	—	—	—
Helium	18.7	19.6	21.0	23.2	27.3	31.2	34.8	38.4	41.8
Hydrogen	8.4	8.8	9.4	10.4	12.1	13.7	15.3	16.9	18.4
Krypton	23.4	25.0	27.4	31.2	38.0	44.2	49.9	55.2	60.2
Methane	10.3	11.0	11.9	13.5	16.3	18.8	21.1	23.3	25.3
Neon	29.8	31.3	33.6	37.0	43.2	48.9	54.3	59.4	64.4
Nitrogen	16.6	17.6	18.9	21.2	25.1	28.6	31.9	34.9	37.8
Nitrous oxide	13.7	14.7	16.1	18.4	22.9	27.0	30.7	34.0	37.0
Oxygen	19.5	20.4	21.8	24.4	29.3	33.7	37.6	41.3	44.7
Steam	9.2	9.7	10.6	12.4	16.2	20.3	24.5	28.6	32.6
Sulphur dioxide	11.6	12.6	14.0	16.4	20.9	25.1	29.0	32.6	36.1
Xenon	21.2	22.8	25.1	28.8	35.7	42.0	47.9	53.4	58.6

For viscosity of gases µPa s is a convenient size of unit. From the kinetic theory the viscosity is expected to be independent of pressure and to vary as the square root of the absolute temperature. The first is true except at very low and at high pressures; the second requires correction. Dividing the kinetic theory expression by a correction factor which is a linear function of the reciprocal of the absolute temperature leads to Sutherland's formula $\mu = KT^{3/2}/(T+C)$, where K and C are constants characteristic of the gas. For higher accuracy a polynomial can be used instead of a linear factor.

(ii) Viscosities Of Some Gases At High Pressure. μ /µPa s (all at 300 K)

Gas	Pressure/MPa				
	2	5	10	20	30
Air	18.7	19.3	20.5	23.7	27.5
Argon	23.3	24.0	25.7	30.5	36.4

Helium	19.9	19.9	20.0	20.1	20.3
Hydrogen	8.98	9.01	9.09	9.31	9.59
Methane	11.6	12.3	14.0	19.2	24.7
Nitrogen.	18.3	18.9	20.1	23.2	26.8
Oxygen	20.9	21.5	22.9	27.1	32.2

(Iii) Viscosity Of Nitrogen At High Pressure. μ /μPa s.

Temperature/K	Pressure/MPa				
	5	10	20	30	50
200	14.6	17.6	26.4	34.9	48.9
250	16.7	18.5	23.1	28.4	38.8
300	18.9	20.1	23.2	26.8	34.4
350	20.9	21.9	24.2	26.9	32.7
400	22.9	23.7	25.5	27.6	32.3
500	26.5	27.1	28.4	29.9	33.2

References

1. Barrer (1943) *Trans. Farad.*, **39**, 48.

2. Bingham and Jackson (1918) *National Bureau of Standards, Bulletin*, **14**, 59.

3. P. W. Bridgman (1926) *Proc. Am. Acad. Arts Sci.*, **61**, 57–99.

4. J. H. Dymond, J. Robertson and J. D. Isdale (1981) *Int. J. Thermophysics*, **2**(2), 133–154; ibid., **2**(3), 223–236.

5. Engineering Science Data (ESDU) Physical Data (1966–83)
6. Chemical Engineering, vol. 3, Viscosity, London.

7. International Critical Tables (1928).

8. G. J. Janz (1968) *NSRDS-NBS 15*, *report*, Natl. Bur. Standards, Washington.

9. Kestin, Sokolov and Wakeham (1978) *J. Phys. Chem. Ref. Data*, 7(3), 941.

10. Landolt-Börnstein (1969) Vol. II, *Properties of Matter in its Aggregated States*, Part 5a, Viscosity and Diffusion, 6th edn, Springer-Verlag, Berlin.

11. Segur and Oberstar (1951) *Ind. Eng. Chem.*, **43**, 2117.
    Smithell (1983) *Metal Reference Book*, 6th edn, Butterworth, London, p. 14.2.

12. R. H. Stokes and R. Mills (1965) *Viscosity of Electrolytes and Related Properties*, Pergamon Press, London.

13. K. J. Young (1980) PhD Thesis, University of Glasgow.

J.T.R.Watson

VISCOSITY OF GASES

$$\mu = A + BT + CT^2, \mu P \text{ (microPoise)}$$

Compound	Formula	A	B, x 10^{-2}	C, x 10^{-6}
Halogens				
Fluorine	F_2	22.09	76.9	-211.6
Chlorine	Cl_2	5.175	45.69	-88.54
Bromine	Br_2	2.153	54.5	-122.2
Iodine	I_2	-17.75	54.71	-99.7
Sulphur Oxides				
Sulphur dioxide	SO_2	-3.793	46.45	-72.76
Sulphur Trioxide	SO_3	4.207	47.12	-68.34
Nitrogen Oxides				
Nitrous Oxide	N_2O	32.28	44.54	-77.08
Nitric Oxide	NO	56.77	48.14	-84.34
Nitrogen Dioxide	NO_2	n. a.		
Carbon Oxides				

APPENDIX V: VISCOSITY TABLES

Carbon Monoxide	CO	32.28	47.47	-96.48
Carbon Dioxide	CO2	25.45	45.49	-86.49
Hydrogen Halides				
Hydrogen Fluoride	HF	-19.21	45.98	-79.96
Hydrogen Chloride	HCl	-9.554	54.45	-96.56
Hydrogen Bromide	HBr	-23.37	74.03	-144.8
Hydrogen Iodide	HI	-17.65	69.77	-136.5
Nitrogen Hydrides				
Ammonia	NH_3	-9.372	38.99	-44.05
Hydrazine	N_2H_4	-17.05	34.01	-47.51
Hydrogen Oxides				
Water	H_2O	-31.89	41.45	-8.272
Hydrogen Peroxide	H_2O_2	5.381	28.98	38.4
Diatomic Gases				
Hydrogen	H_2	21.87	22.2	-37.51
Nitrogen	N_2	30.43	49.89	-109.3
Oxygen	O_2	18.11	66.32	-187.9
Olefins				
Ethylene	C_2H_4	3.586	35.13	-80.55
Propylene	C_3H_6	-5.601	31.88	-62.91
1-Butene	C_4H_8	-8.884	29.59	-57.24
Alkanes				
Methane	CH_4	15.96	34.39	-81.4
Ethane	C_2H_6	5.576	30.64	-53.07
Propane	C_3H_8	4.912	27.12	-38.06
Xylenes				
o - Xylene	$C_6H_4(CH_3)_2$	1.776	21.74	-20.57
m - Xylene	"	-15.27	25.44	-43.43
p - Xylene	"	-13.9	25.57	-44.57
Aromatics				
Benzene	C_6H_6	-15.76	32.45	-72.32
Naphthalene	$C_{10}H_8$	-24.86	27.65	-49.55
Alkyl Aromatics				
Toluene	$C_6H_5CH_3$	-8.421	27.11	-40.18

Ethylbenzene	$C_6H_5C_2H_5$	-14.17	27.37	-47.32
Cumene	$C_6H_5CH(CH_3)_2$	-13.85	25.67	-43.3
Benzene Derivatives				
Chlorobenzene	C_6H_5Cl	-15.08	27.85	0.4464
Aniline	$C_6H_5NH_2$	-14.98	29.03	-1.116
Phenol	C_6H_5OH	-16.41	32	0
Cycloalkanes				
Cyclopropane	C_3H_6	-7.787	34.78	-81.3
Cyclobutane	C_4H_8	-7.558	31.22	-65.69
Cyclopentane	C_5H_{10}	-7.935	28.88	-52.38
Cyclohexane	C_6H_{12}	-4.705	26.32	-44.1
Olefin Monomers				
Isobutylene	C_4H_8	-7.039	31.72	-69.39
Styrene	$C_6H_5CHCH_2$	-5.683	23.68	-32.68
Diolefins				
1,3,Butadiene	C_4H_6	-10.67	34.32	-80.8
Isoprene	C_5H_8	-0.4474	27.56	-49.78
Chloroprene	C_4H_5Cl	-29.93	41.14	-67.86
Organic Oxides				
Ethylene Oxide	C_2H_4O	-7.614	36.27	-66.32
Propylene Oxide	C_3H_7O	-7.72	33.89	-55.94
Butylene Oxide	C_4H_9O	-2.54	28.25	-35.24
Primary Alcohols				
Methanol	CH_3OH	-5.636	34.45	-3.34
Ethanol	C_2H_5OH	1.396	28.48	12.41
n - Propanol	C_3H_7OH	-20.7	31.44	-14.35
n - Butanol	C_4H_9OH	-18.43	28.67	-10.48
Chloromethanes				
Methyl Chloride	CH_3Cl	0.3847	38.2	-54.97
Methylene Chloride	CH_2Cl_2	-4.929	37.72	-53.9
Chloroform	$CHCl_3$	-6.688	37.26	-50.87
Carbon Tetrachloride	CCl_4	5.698	32.73	-40.28

The temperature range covered is from 0 K to 1000 K with average error of 1 to 3%.

APPENDIX V: VISCOSITY TABLES

Source: J. W. Miller, G. R. Schorr and C. L. Yaws; Chemical Engineering, Nov. 22, 1976; pp 153 – 155; McGraw Hill Book Company, New York.

APPENDIX V: VISCOSITY TABLES

APPENDIX V: VISCOSITY TABLES

www.ingramcontent.com/pod-product-compliance
Lightning Source LLC
Chambersburg PA
CBHW020633220526
45464CB00001B/130